T0383045

# Essentials of Engineering Leadership and Innovation

# Essentials of Engineering Leadership and Innovation

Pamela McCauley

CRC Press is an imprint of the
Taylor & Francis Group, an **informa** business

CRC Press
Taylor & Francis Group
6000 Broken Sound Parkway NW, Suite 300
Boca Raton, FL 33487-2742

© 2017 by Taylor & Francis Group, LLC
CRC Press is an imprint of Taylor & Francis Group, an Informa business

No claim to original U.S. Government works

Printed on acid-free paper
Version Date: 20161112

International Standard Book Number-13: 978-1-4398-2011-7 (Hardback)

This book contains information obtained from authentic and highly regarded sources. Reasonable efforts have been made to publish reliable data and information, but the author and publisher cannot assume responsibility for the validity of all materials or the consequences of their use. The authors and publishers have attempted to trace the copyright holders of all material reproduced in this publication and apologize to copyright holders if permission to publish in this form has not been obtained. If any copyright material has not been acknowledged please write and let us know so we may rectify in any future reprint.

Except as permitted under U.S. Copyright Law, no part of this book may be reprinted, reproduced, transmitted, or utilized in any form by any electronic, mechanical, or other means, now known or hereafter invented, including photocopying, microfilming, and recording, or in any information storage or retrieval system, without written permission from the publishers.

For permission to photocopy or use material electronically from this work, please access www.copyright.com (http://www.copyright.com/) or contact the Copyright Clearance Center, Inc. (CCC), 222 Rosewood Drive, Danvers, MA 01923, 978-750-8400. CCC is a not-for-profit organization that provides licenses and registration for a variety of users. For organizations that have been granted a photocopy license by the CCC, a separate system of payment has been arranged.

**Trademark Notice:** Product or corporate names may be trademarks or registered trademarks, and are used only for identification and explanation without intent to infringe.

**Library of Congress Cataloging-in-Publication Data**

Names: McCauley, Pamela, author.
Title: Essentials of engineering leadership and innovation / Pamela McCauley.
Description: Boca Raton, FL : Taylor & Francis Group, CRC Press, 2017. |
Includes bibliographical references and index.
Identifiers: LCCN 2016026613 (print) | LCCN 2016031341 (ebook) | ISBN
9781439820117 (hardback : acid-free paper) | ISBN 9781439820124 ()
Subjects: LCSH: Engineering--Professional guidance. | Leadership.
Classification: LCC TA157 .M341856 2017 (print) | LCC TA157 (ebook) | DDC
620.0068/4--dc23
LC record available at https://lccn.loc.gov/2016026613

**Visit the Taylor & Francis Web site at**
**http://www.taylorandfrancis.com**

**and the CRC Press Web site at**
**http://www.crcpress.com**

Printed and bound in the United States of America by
Edwards Brothers Malloy on sustainably sourced paper

# *Dedication*

*This book is dedicated to the thousands of engineering students I've had the opportunity to teach, mentor, and lead throughout my career in academia, business, and community service. You have made my engineering journey a true joy ride!*

# Contents

# Preface

The global economy, international engineering challenges, dismal economic climate, and growing shortage of the world's natural resources mean that engineers must not only innovate and solve problems but also be integral parts of the holistic process of developing and disseminating the outcomes of technology. A recent engineering meeting defined the outcomes of effective engineering leadership to be "... the process of envisioning, designing, developing, and supporting new products and services to a set of requirements, within budget, and to a schedule with acceptable levels of risk to support the strategic objectives of an organization" [1]. Though other definitions abound, the basic notion is that engineering leadership involves the application of knowledge, ethics, interpersonal skills, and vision for the attainment of an engineering-related goal.

The development of this text is the result of years of applied experiences in engineering leadership and teaching the principles of engineering leadership to students and professionals. Additionally, the interest and outcomes identified in this area were enhanced through a national science research grant.

## Reference


1. Shaw, Wade H. (2003). *Engineering Leadership.* 2003 IEEE Colloquia Tour, April 2003.

# *Acknowledgments*

I would like to express my gratitude to the many people in my personal and professional life who saw me through to the publishing of this book. Thank you to all of my students over the past 23 years at the University of Central Florida and those I interacted with at the Massachusetts Institute of Technology for being my inspiration to put these heartfelt experiences, lessons, and guidance into print.

My sincere gratitude to Drs. Adedeji B. Badiru, Debra Reinhart, and Howard G. Adams, for being outstanding role models and mentors throughout my engineering, academic, and publishing careers.

Thanks to my amazing team mate, Ms. Elizabeth Lee, for her exemplary, consistent, and gold-standard support. This would not have been possible without her priceless input.

To the countless friends, colleagues, and family members who provided help, talked things over, read, wrote, offered comments, allowed me to quote their remarks, and assisted in the editing, proofreading, and design—I offer my sincerest appreciation.

I also would like to thank Senior Editor Cindy Carelli for enabling me to publish this book, and for continuous encouragement and unwavering confidence in my work.

Above all, I want to thank my daughter Annette, son-in-law Jonathan, Maurice and LaFrance McCauley and the rest of my family, who supported and encouraged me throughout my engineering career and the production of this work. It was, at times, a long and difficult journey for all.

I am truly grateful to everyone that has supported me at all levels with this process and throughout my career.

# Author

In the male-dominated fields of STEM, Dr. Pamela McCauley stands out. Her uniqueness stems not only from the color of her dress, the color of her skin, or the height of her heels, but also from the fact that she is an award-winning professor and an expert in ergonomics and biomechanics who has successfully developed and used her leadership style in an array of leadership positions within universities, professional societies, and private industry to create inclusive organizational climates, while recognizing and showing value for the unique contributions of each team member. This has not always been easy.

Such positions, especially in divisive and hurdle-filled environments, have required the use of numerous skills, including strategic plan development and execution, team building, advocacy development, and the establishment of strategic collaborations. She found that by combining these skills and applying principles of scientific research and analysis to leadership in engineering, she could create a pathway for individuals and diverse teams to successfully work together to achieve their collective and individual goals—a pathway to innovation through leadership in engineering.

As a researcher, Dr. McCauley has an impressive record of achievements in addressing critical issues such as predictive models based on fuzzy set theory, human factors in information security, human factors and ergonomics in large-scale disaster management and health care service delivery with international collaborations in Romania, New Zealand, Malawi, and Portugal. She has been successful in consistently securing research funding throughout her career from esteemed agencies such as the National Science Foundation, the Defense Information Systems Agency, the National Aeronautics and Space Administration, as well as from private corporations.

She has authored or coauthored more than 80 technical papers, book chapters, conference proceedings, and the internationally best-selling ergonomics textbook, *Ergonomics: Foundational Principles, Applications, and Technologies* (CRC Press, 2011). Many of her keynotes related to leadership, diversity, innovation, and STEM education draw from her research-based book, *Transforming Your STEM Career Through Leadership and Innovation: Inspiration and Strategies for Women* (Elsevier, 2013), as well as her book on her personal story, *Winners Don't Quit: Today They Call Me Doctor* (IP Publishing, 2003).

In today's global economy, it is critical for engineers to thrive in the worldwide marketplace; to work across disciplines; and to successfully collaborate with individuals from different backgrounds, cultures, and countries. With this book, Dr. McCauley provides a valuable and practical resource to support current and future generations of engineers to be leaders and change makers. Engineering leaders should be prepared to meet the exciting yet formidable challenge of making the world a better place.

# *Introduction: How to use this book*

This book is designed to develop and prepare engineers as leaders to accept the technical and managerial challenges they will face as professionals. The material that is present has been determined to be valuable learning for engineering leaders as we move toward a global society. The text is accompanied by professor's notes, instructor's manual, a sample syllabus, and case studies for implementation in any science technology engineering and mathematics (STEM) education program offering engineering leadership courses. Additionally, this text will provide an approach and application of engineering leadership principles from an engineering education perspective.

At a time when engineering and innovation in technology is of importance in the global community on many fronts, the ideal goal is for this text to encourage engineers and technical professionals to become effective socially conscious leaders and innovators. The text and course material is designed to create an environment of interactive, high-engagement learning that will produce lifelong skills.

We thank the users of this book and welcome their criticisms and suggestions. We hope they will freely provide feedback that nurtures improvement of this book and associated materials.

# chapter one

# A call to leadership

Leadership is an immense ability to compliment the technological expertise of any engineer. What's more, it is the perfect capability if your goal is to get on, get noticed, and get your line of work on the move. Engineers can make a world of difference through their special careers. Primed with technological knowledge, and logical and problem-solving ability, engineers have the knowledge to produce things that make the world a better place.

As the pace of technical change gathers speed and the global economy becomes more interrelated, engineers will be required to become leaders in a broad range of fields including business, administration, law, medicine, government, and community service. Regardless of the field you choose to go into, you will be required to have exceptional communication and leadership skills to completely comprehend and solve the multifaceted problems of our society. Tomorrow's leaders must practice global cooperation, sustainability, innovation, moral values, honesty, judgment, and moral courage.

But what is leadership? Is it different from management? Can it be learned or are we born as leaders? Leadership is difficult to define. It is a feature that I believe all of us can relate to, although it is difficult to express in a wide context in a way that is relevant to all qualified engineers at all levels. The National Society of Professional Engineers (NSPE) has defined leadership relative to engineering as follows:

> In an engineering context, leadership incorporates a number of capabilities which are critical in order to function at a professional level. These capabilities include the ability to assess risk and take initiative, the willingness to make decisions in the face of uncertainty, a sense of urgency and the will to deliver on time in the face of constraints or obstacles, resourcefulness and flexibility, trust and loyalty in a team setting, and the ability to relate to others. Leadership skills are also important to allow engineers later in their careers to help develop and communicate vision for the future and to help shape public policy. These leadership capabilities are essential for the professional practice of engineering and for the protection of public health, safety and welfare. [1]

A leader must identify goals and have the competence to plan the steps required to achieve them. This does not involve trying to comprehend grand corporate objectives. It includes projects and schedules inside your own range of work and goes further than that, as far as you can. Success as a leader involves identifying the proper results needed and knowing the correct steps, which also involves recognizing the wrong steps. Leaders must possess a clear understanding of what it takes to achieve the overall goal efficiently. And that entails doing the job quickly, effectively, on budget and on time.

A leader need not have complete knowledge; it is adequate to have somebody on the team who has the necessary expertise. It is essential for the leader to identify what is required, and where to get it. The extensive knowledge is vital, not the details. Identifying mistakes is frequently the best sign of leadership. Leaders motivate teamwork, without blame. They anticipate results, and look for answers when results are less than expected.

Thorstein Veblen, an American sociologist, in 1921 in his book *The Engineers and the Price System* argued that "… for a technocracy in which the welfare of humanity would be entrusted to the control of the engineers because they alone were competent to understand the complexities of the industrial system and processes and thereby optimize and maximize its output" [2]. In the international community, the United States has been the recognized global leader in engineering leadership for many years; however, this may be changing if effective strategies for implementing leadership throughout engineering education are not successful. Nonetheless, the United States has been considered the undisputed leader in science and technology and according to Richard B. Freeman [3]:

> For the past half century the U.S. has been the world scientific and technological leader and the preeminent market economy. With just 5 percent of the world's population, the U.S. employs nearly one-third of the world's scientific and engineering researchers, accounts for 40 percent of research and development (R&D) spending, publishes 35 percent of science and engineering (S&E) articles, obtains 44 percent of S&E citations, and wins numerous Nobel prizes. [4]

America's technological leadership globally can be traced to their massive investment in the educational sector, especially in the areas of science and technology. These can be deduced from the fact that 17 of the world's top 20 universities are based in the United States [5].

Today, the United States is the foremost entrepreneurial economy for the reason that it applies latest knowledge in more segments than any other economy in the world.

According to Freeman [3],

> Many companies on the technological frontier are American multinationals: IBM, Microsoft, Intel, Dupont and so on. Analysts attribute the country's rapid productivity growth in the 1990s/2000s to the adaptation of new information and communication technologies to production. Scientific and technological preeminence is also critical to the nation's defense, as evidenced by the employment of R&D scientists and engineers in defense-related activities and in the technological dominance of the U.S. military on battlefields.

The United States still offers the world's leading number of scientists and engineers, but East and Southeast Asia countries, most conspicuously China, have been gradually catching up with them. America's lead is distinctive but declining [6]. There are two major reasons for the fear and concern that U.S. leadership in science and technology is ebbing. First and foremost, globalization and the speedy growth of other countries, such as China and India, in science and technology, may possibly make it more and more difficult for the United States to preserve its relative economic lead. Second is the trepidation of the science and technology building blocks inside the United States—science and engineering (S&E) education, infrastructure, and workforce—not being sustained. Assumed insufficiencies include expenditures on research and development (R&D), predominantly on fundamental research; problems with education in S&E; a lack of S&E workers; a growing dependence on foreign workforce; and the declining attractiveness of S&E careers to U.S. citizens.

As a proof of declining U.S. leadership in science and technology, many Organization for Economic Cooperation and Development (OECD) countries, especially China, have begun to catch up with the United States in the area of higher education and in educating S&E specialists. The number of young people going to college has increased rapidly in these OECD countries and in many less developed countries, particularly China [5]:

> Enrollments in college or university per person aged 20–24 and/or the ratio of degrees granted per 24-year-old and in several OECD countries (Australia, New Zealand, Netherlands, Norway, Finland, the United Kingdom and France) exceeded that in the U.S. In 2001–2002, UNCESCO data shows that the U.S. enrolled just 14 percent of tertiary level students less than half the U.S. share 30 years earlier.

In just nine years, from 2003 to 2012, China's high-tech manufacturing sector grew fivefold, an increase that tripled its contributions to global high-tech manufacturing from 8 percent of the market to 24 percent. The United States, by comparison, made up 27 percent of the global total of high-tech manufacturing in 2012 [6].

Nonetheless, if the United States maintains or improves its effectiveness in moving knowledge from the university labs to money-making products, the U.S. comparative advantage in high-technology areas will be sustained longer than would otherwise be the case. The United States should not take for granted its leadership role in science and technology. Helpful guidelines will possibly assist and strengthen its foothold. Below are some strategic guidelines suggested by Titus Galama and James Hosek [7]:

- **Establish a centrally coordinated, independent body to monitor and evaluate U.S. performance in science and technology over the long term:** Comprehensive, objective assessments of U.S. performance in science and technology, performed periodically, are vital to ensuring its health. They can help to inform public debates, identify problems, and guide the development of new legislation. At the same time, they may quell exaggerated claims of the demise or success of U.S. science and technology.
- **Facilitate highly-skilled immigration to allow the United States to continue to benefit from employing foreign S&E workers:** Currently, offshoring of S&E is driven not only by a need to reduce costs but also by an increasing need to gain access to highly skilled labor. If firms cannot fill their S&E positions in the United States, they may decide to offshore or outsource R&D to take advantage of foreign S&E labor pools. High-skilled immigrants, whose numbers have grown much faster than that of the S&E degree holders in the United States, have been a major factor in the fast growth of the S&E workforce. Foreigners thus help to ensure that the benefits of innovation accrue in the United States by allowing innovative activity to remain and expand in the United States.
- **Increase U.S. capacity to interact with science centers abroad and capitalize on the scientific and technological advances being made elsewhere:** Economic strength and global leadership depend on a nation's ability both to absorb and use new technologies and create them. As emerging nations become stronger in R&D, it will become more critical for U.S. researchers to pursue joint ventures, collaborative research, and residences in foreign universities and laboratories to learn about new technology invented elsewhere [7].

## *Engineering leadership*

Put simply, engineers are educated to solve problems. As a group, we have accomplished this in multiple spheres as evidenced by the innovative products, services, and technologies that result from the application of engineering knowledge. The definition of engineering leadership has evolved as the global economy has changed many things including the way organizations react, product development, and team interaction. The common thread is the need to prepare and transform technical professionals into effective leaders and managers. A few other definitions of engineering leadership are as follows:

> Engineering leadership consists of capabilities and values that transform technical people from individual contributors into those who can lead teams to deliver a complex multi-disciplinary product. [8]

Likewise, the definitions for "Engineering Leadership Education" abound. A comprehensive definition is provided by the University of Toronto, Institute for Leadership Education in Engineering:

> Leadership education is about learning how to effectively handle complex, human challenges that often mean the difference between success and failure. Engineers are taught to think analytically and systematically. Leadership skills build on these strengths to make you a more effective engineer. More than just important, they are critical. [9]

Among all of the larger global issues, from engineering, better medicines, and preventing nuclear terror to securing cyberspace, there is still one group our community has underserved—this group happens to be ourselves. A community must be nurtured and prepared to address relevant engineering needs in order for its individuals to reach their full potential and effectively serve others. But according to one recent estimate, about 6 percent of the U.S. workforce is employed in STEM fields while the STEM workforce accounts for more than 50 percent of the nation's sustained economic growth [10]. One key requirement for growth is for those who are educators, leaders, experts, managers, and practitioners to effectively educate and lead newcomers to the field.

Leadership means different things to different groups of people. More specifically, the term *engineering leadership* is a dynamic term having a handful of definitions. At its origin, leadership, of any kind, requires motivation and not solely self-motivation. It is critical for engineering leaders to focus on others, as well, and in particular also inspiring others to perform up to levels beyond what they otherwise would have believed they were capable of, to solve problems and recognize opportunities as they arrive.

We can then build on this foundational definition of leadership by expanding into the traits that make up what the engineering leader is capable of achieving. For example, the MIT Engineering Leadership Program includes traits such as Core Values, Character, Realizing the Vision, Technical Knowledge, and Critical Reasoning. The engineering director of LinkedIn, Greg Arnold, has said that, "Great engineering leaders aren't born, they're made" [11]. This is strong insight, illustrating from his more than 20 years of professional experience that the development of technical skills, people skills, and mastering processes occurs over time in one's career. In other words, there are no shortcuts in this aspect of the growth of an engineering leader, instead it is a process grounded in commitment and perseverance.

Some years ago, Jessica McKellar and a group of friends from MIT started stealthy chat startup Zulip. Less than two years later, it was acquired by Dropbox [12]. And this was not an anomaly. They had done it once before, selling Ksplice to Oracle just as fast. This wild ride has given McKellar a more diverse set of management opportunities than the average engineer ever sees—she's been a team lead, a founder, a technical leader at a massive corporation, and today, is the manager of dozens at a rapidly growing global startup. (She's also a major figure in the Python community.) She says:

> You have to be intentional about working career growth into your broader engineering planning and execution of projects coming down the road. The really difficult thing is that not very many people have a clear sense of what they want from their job, and even when they do, they aren't forthcoming about it with their managers. Good leaders are experts at surfacing this kind of data and making it actionable. [13]

## *What is engineering leadership?*

The many definitions of engineering leadership result from our ever-globalizing economy and society. There is a multitude of opportunities to express leadership skills throughout one's engineering career. In other words, there are no shortcuts in this aspect of the growth of an engineering leader, instead it is a process grounded in commitment and perseverance.

There is a broad spectrum of multinational corporations who have their foundations in technology. Innovation is linked to a country's well-being and for the United States, the Great Recession of 2008 and resultant changes in our economy have had a substantive impact. The International

Labour Organization (ILO) predicted that at least 20 million jobs would have been lost by the end of 2009 due to the crisis—mostly in "construction, real estate, financial services, and the auto sector"—bringing world unemployment above 200 million for the first time [14]. In December 2007, the U.S. unemployment rate was 4.9 percent [15]. By October 2009, the unemployment rate had risen to 10.1 percent [16]. This is why it is so important to establish a foundation for engineering leadership and further to understand what it is and why it matters.

It is more important to remember that engineering leadership can be learned. Although some individuals are naturally more inclined to be led, this does not mean those who have dissimilar personalities should consider themselves as nonleaders; quite the contrary. As with any skill set, leadership within the engineering community can absolutely be learned. It is for this reason that more engineering departments have begun to fortify their curricula with leadership concepts and guidance over and above the traditional foundational knowledge.

A project that reviewed formal engineering leadership programs in academic institutions provided an overview of several international educational endeavors focused on preparing engineering students to be leaders. Today's engineers no longer hold the leadership positions in business and government that were once claimed by their predecessors in the 19th and 20th centuries, in part because neither the profession nor the educational system supporting it have kept pace with the changing nature of both our knowledge-intensive society and the global marketplace. Clearly, new paradigms for engineering education are demanded to:

- Respond to the incredible pace of intellectual change (e.g., from reductionism to complexity, from analysis to synthesis, from disciplinary to multidisciplinary).
- Develop and implement new technologies (e.g., from the microscopic level of info-bio-nano to the macroscopic level of global systems).
- Accommodate a far more holistic approach to addressing social needs and priorities, linking social, economic, environmental, legal, and political considerations with technological design and innovation to reflect in its diversity, quality, and rigor the characteristics necessary to serve a 21st century nation and world [17].
- To compete with talented engineers in other nations with far greater numbers and with far lower wage structures, American engineers must be able to add significantly more value than their counterparts abroad through their greater intellectual span, their capacity to innovate, their entrepreneurial zeal, and their ability to address the grand challenges facing our world.
- To elevate the status of the engineering profession, providing it with the prestige and influence to play the role it must in an increasingly

technology-driven world while creating sufficiently flexible and satisfying career paths to attract a diverse population of outstanding students. Of particular importance is the need to greatly enhance the role of engineers both in influencing policy and popular perceptions and as participants in leadership roles in government and business.

- To take advantage of the fact that the comprehensive nature of American universities provides the opportunity for significantly broadening the educational experience of engineering students, provided that engineering schools, accreditation agencies such as ABET (Accreditation Board for Engineering and Technology), the engineering profession, and the marketplace are willing to embrace such an objective. Essentially, all other learned professions have long ago moved in this direction (law, medicine, business, architecture), requiring a broader understanding within the baccalaureate education as a prerequisite for professional education at the graduate level.
- The globalization of markets requires engineers capable of working in and with different cultures and knowledgeable about global markets. New perspectives are needed in building competitive enterprises as the distinction between competition and collaboration blurs.
- The rapid evolution of high-quality engineering services in developing nations with significantly lower labor costs, such as India, China, and Eastern Europe, raises serious questions about the global viability of the U.S. engineer, who must now produce several times the value-added output to justify wage differentials.
- Both new technologies (e.g., info-bio-nano) and the complex mega systems problems arising in contemporary society require highly interdisciplinary engineering teams characterized by broad intellectual span rather than focused practice within the traditional disciplines. As technological innovation plays an ever more critical role in sustaining the nation's economic prosperity, security, and social well-being, engineering practice will be challenged to shift from traditional problem-solving and design skills toward more innovative solutions imbedded in an array of social, environmental, cultural, and ethical issues.
- There is increasing recognition that leadership in technological innovation is key to the nation's prosperity and security in a hyper-competitive, global, knowledge-driven economy [18].
- Preeminence in technological innovation requires leadership in all aspects of engineering: engineering research to bridge scientific discovery and practical applications; engineering education to give engineers and technologists the skills to create and exploit knowledge and technological innovation; and the engineering profession and practice to translate knowledge into innovative, competitive products and services.

We can look at leadership from two core vantage points. The first is that a strong leader does not exist without strong support. In other words, everyone is an important aspect of the team and leadership is in part, every team member's responsibility. The second is that any effective leader must possess the capacity to lead themselves. Are you able to motivate and encourage yourself? How would you, as an engineering leader, be able to understand others if you cannot first understand, motivate, and "lead" yourself?

## *Different aspects of engineering leadership*

We live in a time of enormous transformation, a progressively more global social order, driven by the exponential development of new information and interlaced collectively by fast changing information and communication technologies. As exciting as this era is, it is equally challenging as growth in the global population creates challenges for worldwide sustainability. In addition, a financial system driven by global knowledge places a premium on high-tech workforce dexterity through outsourcing and offshoring; new prototypes such as open-source software and open-content information challenge conventional free-market beliefs. Additionally, changing geopolitical anxieties that are driven by disparities in wealth and power around the globe, may also contribute to some of the contemporary threats to homeland security by terror campaigns. Nonetheless, it is also a time of remarkable opportunity and hopefulness as new technologies not only advance the human situation but also facilitate the formation and thriving of new communities and communal establishments more proficient at addressing the wants of our world.

To improve the nation's trade and industry output and develop the value of life globally, engineering education in the United States has been looking ahead in recent years to understand and adapt to the remarkable changes taking place in the practice of engineering. The National Academy of Engineering's (NAE) Engineer of 2020 study recommends educating engineers for the future beyond technical abilities to prepare them to succeed in leadership responsibilities in industry, government, and academic career fields. Engineering schools must attract the most excellent and brightest students and be open to new training and teaching advances. With the proper edification and preparation, the engineer of the future will be relied upon to become a leader not just in business but in nonprofit, community, and government areas as well.

There are several promising approaches to such an attempt. For case in point, the Engineer of 2020 study [19] makes use of scenario planning, in which one creates a number of scenarios or stories of probable futures to demonstrate restraining cases at the same time taking benefit of the power of the narrative.

The Engineer of 2020 study is the outcome of the inventiveness of the NAE that endeavored to plan for the future of engineering by asking the question, "What will or should engineering be like in 2020?" [19].

Will it be an evidence of the engineering of today and its precedent growth patterns or will it be essentially different? Most significantly, can the engineering occupation play a role in determining its own future? Can a future be formed where the engineering discipline celebrates the stimulating roles that engineering and engineers play in tackling societal and technological challenges? In what ways can engineers best be trained to be leaders, being able to balance the benefits offered by new technologies with the susceptibilities created by their by-product without conceding the well-being of society and the human race? Will engineering be seen as a groundwork that prepares citizens for an extensive range of innovative career prospects? Will engineering produce products and services that elevate as well as celebrate the variety of all people in our society? No matter what the answers to these questions are, without any misgivings, complex problems and opportunities lie ahead that will call for engineering resolutions and the ability of an ingenious engineering approach.

For the reason that an accurate guess of the future is complicated at least, the Engineering's Engineer of 2020 committee approached its mandate using the method of scenario-based planning. The advantage of this approach was that it eradicated the need to create a consensus view of a single future and offered several possibilities. This method has demonstrated its merit in offering flexible approaches that have the ability to adapt to changing circumstances. Precise scenarios measured in this project were:

- The Next Scientific Revolution
- The Biotechnology Revolution in a Societal Context
- The Natural World Interrupts the Technology Cycle
- Global Conflict or Globalization?

These at times colorful versions only partly capture the dynamic deliberations and debates that took place; nevertheless, they serve to demonstrate and record the thoughts involved in the course of action. Everyone in their own manner informed the discussions about potentials that can outline the tasks that engineering will play in the future.

The executive summary of the *The Engineer of 2020: Visions of Engineering in the New Century* as provided by the NAE as part of their mission to educate the world on issues of science, engineering, and health is outlined below [19]:

- The Next Scientific Revolution scenario offers an optimistic future where change is principally driven by developments in technology. It is assumed that the future will follow a predictable path where technologies that are on the horizon today are developed to a state where they can be used in commercial applications and their role is optimized to the benefit of society. As in the past, engineers will exploit new science to develop technologies that benefit humankind, and in others they will create new technologies de novo that demand new science to fully understand them. The importance of technology continues to grow in society as new developments are commercialized and implemented.

- The Biotechnology Revolution scenario addresses a specific area of S&E that holds great potential but considers a perspective where political and societal implications could intervene in its use. In this version of the future, issues that impact technological change beyond the scope of engineering become significant, as seen in the current debate over the use of transgenic foods. While the role of engineering is still of prime importance, the impact of societal attitudes and politics reminds us that the ultimate use of a new technology and the pace of its adoption are not always a simple matter.
- The Natural World scenario recognizes that events originating beyond human control, such as natural disasters, can still be a determinate in the future. While in this case, the role of future engineers and new technologies will not only be important to speeding the recovery from a disastrous event, but also can help in improving our ability to predict risk and adapt systems to prepare for the possibilities of minimizing impact. For example, there is the likely possibility that computational power will improve such that accurate long-range weather predictions will be possible for relatively small geographic areas. This will allow defensive designs to be developed and customized for local conditions.
- The Globalization scenario examines the influence of global changes, as these can impact the future through conflict or, more broadly, through globalization. Engineering is particularly sensitive to such issues because it speaks through an international language of mathematics, science, and technology. Today's environment, with issues related to terrorism and job outsourcing, illustrates why this scenario is useful to consider in planning for the future [19].

After all the assessments and discussions, the authors conclude that the next several decades will present more opportunities for engineers, with exhilarating potentials expected from nanotechnology, bioengineering, and information technology. Other engineering applications, such as transgenic food, technologies that have an effect on individual privacy, and nuclear technologies, raise multifarious social and moral challenges. Engineers of the future must be ready to assist the public reflect on and resolve these problems alongside other challenges that will crop up from new global rivalry, necessitating considerate and determined exploits if engineering in the United States is to maintain its effervescence and power.

## *Elements of engineering leadership*

### *Engineer your own success (book review)*

Every young engineer's first leadership role should be that of managing his or her own career. The book, appropriately titled, *Engineer Your Own Success* [20] is a guide to improving competence and performance in any engineering field. It imparts important organization tips, communication advice, networking tactics, and practical support for preparing for the Practice of Engineering (PE) exam—every needed skill for success. Authored by Anthony Fasano, the book provides highly useful and practical strategies for climbing the career ladder as an engineering professional. Key elements of this guide are as follows:

1. **Obtain Credentials that Will Help You Reach Your Goals**
   This chapter presents some suggestions and approaches on how to go about acquiring qualifications that can put a person in a place to be successful. The writer describes the grounding strategy for exams that has worked for him every time and emphasizes the distinction between patience and procrastination. The chapter further provides instructions for approaching the PE exam in the United States and sums up the credentialing process for 11 other countries with large engineering populations. A Master of Business Administration (MBA) or engineering management certificate may give an engineer something very vital that a graduate-level engineering certificate may not have, which is flexibility.

   The chapter also stresses on the need to draw attention to the awards received during the career and the necessity to take advantage of reimbursement programs offered by corporations for professional growth activities.

   There are certain credentials that your goals will not be possible without. For example, let's say you want to achieve partnership status in an engineering company, and the company requires a master's-level degree and/or a professional engineering license; well, no matter how great you are at business development, you will have to gain one of those credentials to achieve your goal. While all of your skills and business accomplishments can walk you through that door to partner, having those credentials is the key that must first unlock the door before you can walk (or run) through it.

2. **Find and Become a Mentor**
   This chapter provides suggestions for helping you discover the right mentor and also becoming one. It is not all the time easy to find a mentor; if you find one, take full advantage of each minute that you have to spend with this knowledgeable professional who has committed to help you achieve your goals. Although having a mentor is a significant part of advancing your career, being one yourself is just as vital. Giving to others will facilitate in you growing immensely, both personally and professionally. If you have made up your mind to become a mentor, the next step will be to find a mentee to work with. Discretion and responsibility are key parts in a successful relationship. As a mentor, it will be your responsibility to provide some guiding principles for how the real relationship will work.

   This chapter offers some recommendations for carefully or carefully transitioning from a mentoring relationship. The guidelines will list some of the things not to do for both the mentee and mentor, an instance of this condition is an official mentoring program that allows organizations to create and cultivate such relationships by matching more qualified employees (mentors) with less-skilled employees (mentees) to meet specific agency objectives while helping individuals in the mentoring relationship to identify and develop their own talents.

3. **Become a Great Communicator**
   Engineering professionals expend a good quality part of their careers working in a teamwork environment. The capability to converse with the clients or coworkers will directly impact your career success as an engineer. One of the challenges you will face will be how to communicate technical information to nontechnical people. This part of this book presents strategies on how to improve communication skills and increase your confidence and helps you to become an effectual communicator. Sincerity and reliability are completely critical for achievement in all areas of life. The chapter emphasizes how significant public speaking skills are for you as an engineer and suggests joining Toastmasters to start developing communication skills. The chapter also offers some specific advice for improving your public speaking skills.

4. **The Ability to Network**
   As an engineer, networking is a word that you will hear frequently in your career as an engineering professional. Lots of people are of the opinion that networking is going out to social outings, collecting business cards, and attempting to secure clients on the spot. That is not what it is. In fact, networking or building relationships is in effect characterized by what a person does after that first meeting. In this chapter, you will learn how to conquer individual obstacles

to networking and building strong and lasting relationships. The chapter also offers great tips on how to build strong relationships, as well as some secrets that the author learned through personal experience. The best method to build relationships is to focus on building individual associations or friendships.

5. **Stay Focused, Organized, Productive, and Stress-Free**
The three key ways to efficiently utilizing time are being organized in every effort, staying focused and fruitful at all times, and avoiding anxiety and worry at all costs. If these three rules are followed, a well-balanced, low-stress, fulfilling life, both personally and proficiently, can be created. Being organized impacts you in more ways than you think. A lot of professionals struggle with staying structured in the workplace. The capability to focus straightforwardly impacts on how productive you are in every sector of your work and life. Pressure and worry are two feelings that can overcome anyone. One of the most significant aspects of attaining work–family balance is flexibility. Built-in work flexibility gives more freedom to manage time. This chapter lays out a few steps to help define what balance means to a person and to achieve it in his or her career [20].

## *Capabilities of engineering leaders*

Engineering leadership consists of capabilities and values that transform technical people from individual contributors into those who can lead teams to *deliver* a complex multi-disciplinary product. We assert that leadership is a process and that there is a two-way relationship between the leader and the team. Leaders inspire and influence teams to accomplish things that they otherwise would not have done on their own.

### *What is engineering management?*

Engineering management is a dedicated form of management that is concerned with the use of engineering principles in business practices. Engineering management is an occupation that brings together the technical problem-solving know-how of engineering and the organizational, managerial, and forecast abilities of management in order to oversee complex enterprises from conception to completion [21]. A Master of Science in Engineering Management (MSEM, or MS in Engineering Management) is very often equated to a MBA for experts seeking a graduate degree as a qualifying credential for a career in engineering management [22].

Example areas of engineering are product growth, manufacturing, construction, design engineering, industrial engineering, technology, production, or any other field that employs workers who carry out engineering functions.

Successful engineering managers usually need training and know-how in business and engineering. Technically inept managers tend to be deprived of support by their technical team, and noncommercial managers tend to lack commercial acumen to deliver in a market economy. For the most part, engineering managers manage engineers who are driven by nonentrepreneurial thinking and thus require the necessary people skills to coach, mentor, and motivate technical experts.

## *How leadership is different from management*

Just like every engineer needs organizational competencies to get work done, a completely different set of skills are also important—leadership skills. Leadership is completely interpersonal but at a different level than the interpersonal competencies that are described as "management."

It's a common fallacy that leadership and management are one and the same. While managers can be team leaders and vice versa, there is a disparity between the skill set necessary for management and leadership.

Knowing and understanding the disparity between the two roles will make leaders and managers more successful in building their business.

In fact, the dissimilarity between leadership and management is risk. Leaders are at ease taking risks that can bring enormous rewards. As a consequence, they frequently make complex decisions for the good of the business. These can range from reorganizing the organization to making new product offerings. Meanwhile, managers are often less comfortable with making those decisions themselves. Instead of making decisions hurriedly, managers tend to be contented asking lots of questions as they nurture actionable plans to make leaders' choice a reality.

As leaders and managers work together, they need to be aware of the difference between their functions. This understanding will cede improved results for the organization.

Management is a set of well-known procedures, such as planning, budgeting, structuring jobs, staffing jobs, measuring performance, and problem-solving, which help an organization to inevitably do what it knows how to do well. Management aids you to create products and services as you have guaranteed, of consistent quality, on budget, day after day, week after week. In organizations of any size and intricacy, this is an extremely difficult task. We continually underrate how multifaceted this task really is, particularly if we are not in senior management jobs. So, management is essential, but it's not leadership.

Leadership is completely different. It is concerned with taking a business into the future, looking for opportunities that are coming at it quicker and faster and successfully taking advantages of those opportunities. Leadership is concerned about vision, about people buying in, about empowerment, and most of all about creating useful change. Leadership is not about attribute, it's about performance. Furthermore, in an ever-faster-moving world, leadership is more and more needed from more and more people, no matter where they are in a chain of command. The conception that a few astonishing people at the top can supply all the leadership required today is preposterous and it's a formula for failure.

Some people still bicker that we must swap management with leadership. This is noticeably not so: they provide different, yet indispensable functions. We need outstanding management and we need more outstanding leadership. We need to be able to make our multifarious organizations dependable and proficient. We need them to jump into the right future at an accelerated pace, no matter the dimension of the changes needed, to make that happen.

There are very few organizations these days that have adequate leadership. Until we face this problem, understanding precisely what the problem is, we're never going to resolve it. Unless we emphasize that we're not referring to management when we speak of leadership, all we will attempt to do when we do require more leadership is work harder to manage. At some definite point, we will end up with overmanaged and under-led organizations, which are more and more at risk in a fast-moving world.

## *Engineering leadership skills*

The most excellent engineers are regularly highly efficient managers and leaders. If you are motivated and plan to advance your career, attaining these skills is critical. For small production companies, strong inspiring leadership and effectual management is the surest means to get the very best out of people, assets, and limited resources.

Perhaps for you it's more personal. You may possibly have been promoted to a management role because of your achievement as a technological expert. This is great, apart from the fact that lots of your time is now spent supervising teams and dealing with people issues, with astonishing small deep technical work. What you now require are the new skills to endure and succeed in this brave new world.

On the ground, the functions of engineers and technicians naturally comprise management-related duties. Engineering employees find themselves accountable for leading teams and running projects, and at the same time being accountable for budgets and assets as well. They work with suppliers and customers, in addition to being in frequent contact with other tasks within the business.

Nevertheless, the truth is that good engineering and operations managers are in limited supply. Engineering frequently deals with details, while management is regularly more subjective and vague—so perhaps it's not unanticipated.

Bearing in mind the above, strangely enough the availability of good quality information to assist engineering managers direct and succeed is far less accessible than comprehensive technical training. But if engineers are to truthfully appreciate their potential and progress their careers, they need to reflect on and act upon basic management skills. In addition, if engineering is to be well represented at every level of the company, all the way up to the board, engineers need to promote their knowledge in this area. Anywhere you are in your career, you must make every effort to get hold of and practice these managerial skills.

## *A promise to future generations—A new oath for graduating engineers*

Envision a world where engineers were bound to revere the rights of future generations. At the moment, we face some stern challenges. Worldwide problems of global warming, climate change, and the thinning supplies of inexpensive carbon-based energy threaten our biosphere and our developed economies. These challenges not only compromise our ability but also the ability of upcoming generations to meet their needs, to accomplish their dreams, and to decide their destinies.

Resolving these global problems indicate that we need to work together to make new sustainable and dependable decisions, and we have to begin making them now. Even though change is taking place, according to William D. Ruckelshaus, former head of the U.S. Environmental Protection Agency, bringing about the required paradigm shift toward more sustainable behavior will be a momentous task. He believes that [23]:

> Moving nations and people in the direction of sustainability would be a modification of society comparable in scale to only two other changes: The Agricultural Revolution of the late Neolithic and the Industrial Revolution of the past two centuries. Those revolutions were gradual, spontaneous, and largely unconscious. This one will have to be a fully conscious operation, guided by the best foresight that science can provide. If we actually do it, the undertaking will be absolutely unique to humanity's stay on earth. [23]

As consciousness develops, the need for transformation becomes more compelling. The enlightened no longer argues whether change is compulsory but asks how scientists and engineers should respond so that systematic and practical change is brought about.

The new oaths by newly graduating engineering students are outlined here [24]:

- **Article I:** Each generation has the right to inherit a healthy earth where they can develop their culture and social bonds as a member of one intergenerational family, and each generation has a corresponding responsibility to accord a similar right to future generations.
- **Article II:** All generations, sharing in the estate and heritage of the earth, have a duty as trustees for future generations to use resources with forethought and responsibility, to honor life on earth, and to foster human freedom.
- **Article III:** In fulfilling the duty owed to future generations, it is the paramount responsibility of each generation to be prudent and constantly vigilant to ensure that biodiversity and the balance of nature are respected.
- **Article IV:** All appropriate measures shall be taken to ensure that the rights of future generations are protected and not sacrificed for the expedience and convenience of the present generation.
- **Article V:** The rights of future generations have a claim on the conscience of all peoples. To develop a culture that promotes respect for individuals, society, and the environment, every person is challenged to imaginatively implement these principles as if in the very presence of those future generations whose rights we seek to perpetuate.

Our culture today is faced with a number of unbelievable challenges, and in order to pilot the long road ahead we have to start executing the principles and standards of the promise to future generations. As engineers, we must discontinue settling for the status quo, we must stand up, speak up, and lead. Can you envision the human race where all engineers promote and concur to be bound by such a promise?

## *Uncovering leadership opportunities*

What is leadership? It has been delineated in scores of different ways by many different people. A number of definitions stress the significance of having a vision, others think that spirit and personality are key. Several meanings make leadership appear likely just for truly motivated individuals. Nonetheless, our definition is that the real meaning of leadership is very simple, so simple in fact that anyone can be a leader. In fact, leadership is all around you right now, in restrained ways. The bold girl in your class who calls out to the loud students at the back to be quiet; the boy who assembles several perplexed peers to create a study group. These are just normal people doing little things to make helpful changes in their lives and in the lives of those around them. Nevertheless, these little steps can ultimately lead to something superior that can have a noteworthy effect on an organization, a country, or the world.

While we primarily focus on students who have never before put themselves into leadership positions, the methods that we recommend can be utilized by anybody, whether an experienced team leader or a trainee, before and after they complete a task, work with a new team, or take on a new role.

You may be baffled and wondering "Why me? I am just a student besieged with course work and way too numerous tests. How can I possibly find sufficient time to even imagine becoming a leader?" All the same, what if we demonstrate to you that developing your leadership skills may possibly in fact help you execute better, learn better, and feel better? Applying your leadership capacity in team settings can produce a hard-working and pleasant atmosphere.

Companies and industries look to employ people with leadership skills. For the reason that graduates are normally assumed to have the same technological knowledge, it is these less tangible skills that will give you an advantage. Even once you have been hired, some companies present opportunities to advance your leadership skills. But why start then? There are numerous opportunities in which to completely learn leadership, such as sports teams, musical groups, volunteer organizations, and student government. Take advantage of these opportunities to put into practice your leadership skills in a supportive environment. But where do you start developing your leadership potential? How do you go about finding leadership opportunities?

According to the *Engineering Leadership Review* [25], leadership can be illustrated as a four-step process of self-reflecting, getting started, leadership in action, and leadership learning.

## *Self-reflection*

Prior to getting involved, it is imperative to spend some time getting to know yourself. Recognizing what you want to obtain from your experience can help you find an organization in which to get involved. In addition, as a leader, it is very useful to know how you will respond in certain circumstances and discover how to keep away from reacting in a self-protective manner. At this stage, you can recognize your strengths and weaknesses: what you are good at, and what do you do with difficulty. Endeavor to understand your personal enthusiasm, principles, and well-being.

## *Getting started*

When choosing an activity, there are a couple of things to consider. Being occupied in school-related activities indicates you will most probably be meeting with the group on campus, while participation in another group of people may entail meeting at a local church, school, or community center. If this is your initial effort at getting involved, reflect on joining clubs that are close to home. This may persuade you to contribute more often. In a similar way, if you are disinclined, choosing an association that your friends are part of can help you feel more at ease in your new position. Nevertheless, if you find a prospect that matches your goals and ideals, don't be afraid to travel the additional distance to attend meetings.

### *Leadership in action*

There are numerous skills that are helpful in leadership roles. You will study some of them in classes, such as conflict resolution and engineering strategies and practice. Others, such as networking, you may intrinsically possess or pick up from watching others in diverse group circumstances. Every one of these skills can be learned, and the most excellent way to become more proficient with these skills is in the course of practice.

### *Leadership learning*

In 1963, President John F. Kennedy said, "Leadership and learning are indispensable to each other" [26]. Immediately you become involved, you may perhaps want to complement your implied learning with recognized training or theory. As engineering students, you are already on track. Engineering Strategies and Practice I & II give you a savor of major leadership skills, such as team management. The leaders of tomorrow infusion lectures also build on what you have already learned.

### *Conclusion*

Just as engineering needs a toolbox of math, science, logic, and design strategies, leadership also requires a unique set of skills that can be built on in the course of practice and experience. By discovering participation opportunities that encourage you to make a difference, you can identify your own potential as a leader. I believe that everybody has the potential to affect their society, school, or peers in an affirmative way.

## *Leadership impact*

- **Leadership for Adopting Social Change**
  Leadership is influential in achieving communal change. In the entire history, whether it was for putting an end to social norms, conquering social evils, or modernizing history, social change has been unfeasible without the right type of leadership. When it comes

to rallying the masses, igniting enthusiasm in people toward a general goal, and inspiring people to do something toward the said general goal, it isn't possible to bring together people and motivate them to action without leadership.

- **Leadership for a Positive and Content Society**
  It is fascinating to note that one individual or a small group of people has the clout to manipulate how millions feel. A society that lacks capable leaders is perpetually thrown into discontent at a small scale and chaos and lawlessness at a larger scale. A leader who is excellent at what he or she does, is capable of keeping people motivated and stimulated, works for the betterment of society and not just for his or her own personal benefits, and is successful in creating a constructive and happy society is revered by the people.
- **Leadership for Improved Professional Performance**
  It is quite extraordinary that even when leadership is effectual in the social, but not business context, it has an impact on people's professional lives. When a society is led by a commanding, optimistic, and progressive leader, one of the most important areas of focus is people's professional growth. It goes without saying that the professional growth is essential for economic growth and no society can do well without economic stability. Consequently, excellent leaders are those that take all dynamics into deliberation, even if their role is supposedly limited to one task.
- **Leadership for a Strengthened Identity**
  A good number people fail to value how an ordinary leader is time and again the face of the society and its symbol as well. When people select a leader they are proud of, or they are positioned under the care of a leader who does an excellent job, there is a sense of satisfaction and identification with the individual that furthermore binds the society together. An effectual leader is one that the populace of a society is pleased to call their own, and in turn, the leader reciprocates by bringing the society together and giving them a widespread, positive personality that the people are all pleased to have.

## Applied engineering leadership

Applied engineering leadership goes beyond the classroom to the individuals who are currently both engineers and engineering leaders. It pertains to the daily, real-world experience in the trenches, planning, executing, evaluating, and accomplishing the given technical task while also maintaining the highest levels of responsibility to social, environmental, and stakeholder groups involved. Some skill sets such as conflict resolution may be learned in class, whereas others will be honed and perfected through repetitions during one's career. Four other important such skills that are learned and then perfected through practice are as follows:

1. **Vision**

   True leaders have a vision, that is, they have a potential to view the present as it is and to invent a future arising out of the present. A leader with a vision can foresee the future and can remain in the present. A vision is an end toward which a leader can spend and direct his or her energy and resources. Leaders share a dream and a path which the employees want to share and follow. Leadership vision is not restricted to an organization's written mission statement and vision statement. It is well demonstrated in the actions, beliefs, and values of an organization's leaders.

   The saying, "If there is no vision, the people perish" [27], is relevant both in business as well as in life. Leaders who lack vision cannot achieve anything in life and they work in an average and repetitive manner. Vision is not a fantasy for leaders, rather it is a reality that is yet to come into practice. Consequently, to have a vision, a leader should bring to bear special additional efforts and have immense self-confidence and commitment to recognize the vision. A vision works as an inner strength, a resource that pushes a leader to act on his or her goals. It presents a leader with a purpose. The reliable existence of a vision makes a leader progressive notwithstanding a variety of adversities and obstacles. Vision is a bond that binds different individuals into a team with a shared goal.

   Acknowledgment of a leader's vision by an organization's employees is very crucial as it makes the workers aware of what the organization is striving to accomplish. Vision has the power to move workers out of a boring work life and to position them for a new demanding and vibrant work. Vision must be:

   a. Rational
   b. Reasonable
   c. Innovative
   d. Credible
   e. Clear
   f. Motivating and stimulating
   g. Challenging
   h. Reflective of organizational ideas, values, and culture
   i. Concrete

   It is the duty of the leader to mold, interpret, communicate, and stand for the vision. Vision is a representation of what a leader desires his organization to be in the long term.

2. **Effective Team Management**
   Team management is the most commonly thought of skill when considering how to be a leader. A leader helps his or her team to work in harmony; a few indicators of effective team management include level of productivity, team morale, appropriate communication, and quality of work produced. They should be able to produce more work as a group than individually and separately. Team management also takes into consideration how each individual fits into a team, including what their strengths are and what types of projects they enjoy and perhaps at which they would excel. By playing to each person's strengths, it becomes possible to compose a more creative, efficient, and effective team.

   There are a number of responsibilities that you cannot handle alone. People need to come together and talk about things among themselves and work collectively toward the attainment of a common purpose. The persons forming a team can and should be diverse as this has been shown to be beneficial in team performance. However, this diverse team must have a common goal and shared vision for achieving the objectives. All teams are formed to accomplish a predefined goal, and it is the duty of each and every member to put in his or her level best to achieve the allotted task within the specific time frame. The team members have to complement each other and come to each other's aid as and when necessary. Individual performances do not matter much in a team, rather all persons should make every effort to work in harmony.

3. **Motivation and Inspiration**
   Motivation is a tool employed to drive individuals toward a common goal. Although motivation itself is a singular goal, there are a few ways to go about it:
   a. **Achievement:** by delegating projects to provide a collective sense of ownership, team members are offered opportunities to grow and derive satisfaction that encourages them to perform consistently at their highest level.
   b. **Recognition:** team members who feel their efforts are not put to waste will work harder.
   c. **Providing incentives:** a reason for team members to work beyond their comfort zones.
   d. **Project involvement:** including all team members such that they are enthusiastic about the project and are further driven to reach their goals.

   Motivation in management is portrayed as ways in which managers encourage productivity in their staff. A lot of people are confused about "happy" staff and "motivated" staff. These may possibly be related, but motivation in reality describes the level of desire workers feel to perform, despite the level of their happiness. Workers who are sufficiently motivated to perform will be more dynamic, more occupied, and feel more invested in their work. When employees experience these feelings, it helps them perform better, and in so doing help their managers as well, in becoming more successful.

It is the job of the managers to motivate staff to do their jobs well. So how does a manager do this? The answer is management by motivation, the practice through which managers give confidence to employees to make them productive and effective.

Imagine what you may experience in retail scenery when a motivated cashier is doling out your transaction. This type of cashier will:

a. Be welcoming, creating an enjoyable transaction that makes you more prone to return
b. Process your transaction swiftly, which by implication means that the store can service more customers
c. Propose an additional item you would like to buy, raising more sales for the store

In short, this employee is industrious and delivers a high-quality output.

4. **How to Motivate Employees**

There are numerous ways to inspire workers. Managers who want to inspire more efficiency should work to make sure that employees:

a. Consider that the work they do has meaning or significance
b. Trust that good work is rewarded
c. Believe that they are treated reasonably

5. **Quality Decision-Making**

We make decisions every day, from the ordinary to the life-changing. Nonetheless, the role of decision-making in leadership affects the many, not just the few. The decisions you make as an engineering leader can put your team on the right path for a project. Decisions can be made through voting (useful for topics of lesser importance that need a quick resolution), consensus (mutual team agreement on one solution acceptable to all), and individual decision-making (one individual is ascertained to be the expert on a given issue and is delegated the responsibility for providing the solution).

All organizations need to make choices at one point or the other as part of the administrative process. Decisions are made in the paramount interest of the organization. For that matter, decisions made by the organization are to improve the way forward. Be it tactical decisions, business activities, or HR matters, the processes of making decisions is multifaceted and involves professionals of diverse genre. While small organizations involve all levels of managers, multifaceted organizations mostly depend on a team of professionals particularly trained to formulate all types of decisions. But bear in mind, such a body on its own cannot come out with ultimate decisions. Here the decision making process is a combination of collective input and counseling or communication with stakeholders. The procedure, on the whole, has its advantages and disadvantages would by and large originate results and consequences in the organizations' general development and prospects.

As a matter of fact, the ability to take decisive decisions is one of the several traits that every manager must have, be it at the top level, middle level, or entry level. Decisive decision does not mean choices are made without thought but rather once the decision is made, there is a commitment to it. This type of decision making is an essential quality and if not natural to a leader it may be an acquired skill in the professional development process.

6. **Technical Competence**

To be successful at any undertaking, you should be conversant and have the skills needed to successfully complete your assignments. To look deeper into what it means to be technically competent, you should have the ability:

a. To comprehend and describe the technical situation
b. To offer instruction on the tasks included within the project assignment
c. To give specifics on how to accomplish the tasks
d. To explain the expected result of the different tasks assigned within the project

Technology is a phrase that is regularly used in the business world. It is a term regularly related to science. But there is a noteworthy difference between the two. Science includes the result of fundamental academic studies while technology infers to the pertinent application of science. This is an important distinction to make as science generally contributes to the development of technology. This may include the inclusion of multiple areas of science to create a single technology.

The importance of technology to a business is to be found in the reality that technology can present a competitive advantage. As a result, technology can be measured as an asset of tactical importance. In addition, it can be said that an organization's ability to hold onto and obtain the benefit of technology can be a symbol of core proficiency.

## *Technology competence features*

Technology proficiency can develop an organization's product portfolio. This is achievable in diverse ways as listed below:

- **New Functions**
A new product can be built which allows the client to carry out actions that were earlier not possible or else tremendously hard. For example, reflect on the growth of cellular phone technology. It allows customers to easily communicate around the globe. There are clients who are prepared to compensate high values to obtain products having the newest technology. These products are anticipated to be strikingly original needing excellent investments in new technology.

- **New Features**
An obtainable product can be changed to convert it into a more useful product while the basic efficacy continues to be the same. For case in point, take the examples of mobile cell phones having cameras. Organizations continually search for improvements to make their products unique from those of their rivals. Even if such improvements are small, over a length of time, these can be merged to indicate a key progress in technology.

- **Superior Dependability**
Amid technology mounting into a more complex stage, product reliability turns out to be a main feature in product segregation. For example, enhanced utilization of precise integrated circuits can result in ease of product assembly. Improvement in designs and various techniques of production will focus on performance and value.

- **Reduced Costs**
In the product maturity cycle or as newer generations of a technology are released this may lead to a reduction in costs for development and ultimately a lower cost for the customer. For example, reflect on the use of specialized integrated circuits as referred to earlier. They are expensive to envisage; however, in mass production, they present enormous cost benefits over separate parts. They can offer a wonderful development to the business that can manage this technology.

  Technology is one of the basic reasons for the continuation of a product life cycle. When the technology is in a new phase, improvement is fast and product performance rises quickly. As the technology becomes recognized, the pace of alteration of functioning becomes more stable in view of the fact that the technological threshold has been accomplished. At a definite point, a different new technology is developed and incorporated into the product. The cycle will begin all over again.

## Strategies to develop as an application ready engineering leader

Leadership growth involves cultivating, encouraging, developing, and mentoring potential leaders. In several organizations, the human resource department in arrangement with higher management recognizes would-be leaders or fast trackers who are skilled and make certain that they are motivated and pushed to offer their best. This type of mentoring and training of future leaders happen through organizational dedication to their growth that comprise sending them to specific training programs and making them attend targeted seminars with the express intention of making certain that these potential leaders get all the support and strategies to prepare them for higher-level positions.

It needs to be considered that leadership growth is not merely about the organizational need for grooming potential leaders but in addition it has to do with the candidates themselves showing an inclination and the propensity to be prepared for leadership.

The emphasis here is that leadership development is a two-way procedure that is symbiotic with both the organization and the candidate actively demonstrating an interest in leadership development that supports the individual employee and organization goals. Only when both sides are enthusiastic on assisting each other grow can proper organizational development take place.

Leadership growth is as much about personality as it is about knowledge and this is where the responsibility of a mentor emerges. Existing higher management leaders can assist would-be leaders carry out improved decisions and offer instructions and guidelines on how the business world functions so that they have a plan regarding reacting to tricky and difficult issues.

In conclusion, leadership development involves dedication, endurance, and skillfulness. Unless the prospective leaders are committed to continuing with the organization for a comprehensive period of time, there is no point in grooming them. Despite the fact that lots of organizations have moved away from making the workers sign bonds, they nonetheless groom merely those workers who have been with them for some time and who, in their assessment, intend to stick with the company for a longer period.

Tools and instruments that may be used for leadership development include:

1. **Peer-to-Peer Learning**
   Peer learning is not a single, undifferentiated educational plan. It encompasses an expansive sweep of activities. For instance, researchers from the University of Ulster identified 10 different

models of peer learning [28]. These ranged from the conventional proctor model, in which senior students teach junior students, to the more inventive learning cells, in which students in the same year form partnerships to aid each other with both course content and private concerns. Other models comprised discussion seminars, personal study groups, counseling, peer-assessment method, joint project or laboratory work, place of work mentoring, and neighborhood activities.

The term *peer learning* nevertheless remains conceptual. The sense in which we use it here suggests a two-way, give-and-take learning activity. Peer learning should be equally favorable and entail the sharing of knowledge, thoughts, and know-how between the participants. It can be described as a way of moving beyond independent to interdependent or mutual learning [29].

Carsrud et al. [30] describes an example of a stand-in teaching method in which doctoral students oversee undergraduate psychology students in conducting research projects. One of the most important objectives of this program is to support highly motivated and well-prepared students to become attracted in pursuing research through skill development and first-hand exposure to research. The undergraduates worked closely with the graduate students in designing and implementing the research, and they were required to produce a professional-style report at the end of the study. The program was considered a success, based on participants' self-reports.

2. **Practitioner-to-Practitioner Arrangements**

All practitioners have a professional responsibility to ensure service delivery that is consistently available, appropriately supervised and delivered. This means implementing plans to address coverage needs during holidays, vacations and illness of the practitioner. If this is not done there likelihood for a decline in the value of the practice or even its collapse is possible.

It is consequently essential for a sole practitioner to deal with these problems and difficulties, if possible when he or she first enters into practice, and to make appropriate provisions to each of the following situations to allow the practice to be carried on with the least interruption. An instance of a practitioner-to-practitioner arrangement is the arrangement between a qualified midwife/qualified nurse practitioner and a medical practitioner that must provide for:

a. Discussion with a specified medical practitioner
b. Transfer of a patient to a particular medical practitioner
c. Reassigning of the patient's care to a specific medical practitioner, as clinically appropriate, to ensure safe, high-quality health care

3. **Access to Best Practice and Lessons Learned in Other Countries**
   a. What is a best practice?
      A *best practice* can be referred to as a procedure, practice, or system known in public and private organizations that perform remarkably well and is broadly recognized as improving the performance and effectiveness of organizations in particular areas. Effectively identifying and applying best practices can trim down business expenses and develop organizational competence.
   b. What is a lesson learned?
      A *lesson learned* can be referred to as the experience gotten during a project. These training come from working with or resolving real-world problems. Lessons learned record recognized problems and how to resolve them. Assembling and disseminating lessons learned help to eradicate the occurrence of similar problems in future projects. Lessons learned on average are negative with respect to recognizing procedures, practice, or systems to steer clear of in particular circumstances. Lessons learned are positive with respect to recognizing of solutions to challenges when they occur.

4. **Twinning**

> A twinning is the coming together of two communities seeking, in this way, to take action with a European perspective and with the aim of facing their problems and developing between themselves closer and closer ties of friendship. [31]

A good twinning partnership can fetch several benefits to a society and or a city. By bringing people together from different sections of Europe, it gives a chance to share problems, swap views, and comprehend different viewpoints on any issue where there is a common interest or concern.

There are numerous examples of good practices in twinning which cover an extensive variety of issues including but not limited to sustainable development, art and culture, local public services, young people, local economic development, and citizenship. It symbolizes a long-standing obligation between the partners, not a temporary project affiliation. It should always be able to endure changes in political leadership and short-term problems of one or both partners, and assist each other in times of need, for example, a major flood.

5. **Mentoring**
   Mentoring is a process of building official interactions among junior and senior members of an organization; in some cases, mentoring can also take place between peers. In other words, it is a process of building relationships connecting more knowledgeable members of an organization and the less knowledgeable ones for transmitting knowledge and skills. These relations are developed with the intention of developing career roles. For instance, training, funding, protection to peers, challenging homework, introduction to essential associates and resources are sure ways in which mentoring may happen.

6. **Coaching**
   According to Renton [32], coaching is training or development in which a person called a *coach* supports a learner in achieving a specific personal or professional goal. The learner is sometimes called a *coachee*. Occasionally, *coaching* may mean an informal relationship between two people, of whom one has more experience and expertise than the other and offers advice and guidance as the latter learns; but coaching differs from mentoring in focusing on specific tasks or objectives, as opposed to general goals or overall development.

   Business coaching is an example of human resource development. It offers helpful support, response, and advice on an individual or group basis to develop personal efficiency in the business setting. Business coaching is also referred to as executive coaching, corporate coaching, or leadership coaching [33].

   Coaches assist their client's progress toward precise professional goals. These include career change, interpersonal and specialized communication, performance management, organizational efficacy, managing career and personal changes, increasing executive presence, enhancing strategic thinking, dealing effectively with conflict, and building an effective team within an organization. An industrial organizational psychologist is one example of an executive coach [34].

7. **Access to Knowledge and Information**
   Information is the fuel that runs the world economy. The information know-how and interrelated industries function on the foundation of knowledge as the medium of input and output. Truly, information is wealth and whoever has the expertise has the power.

   In this age of information, how organizations handle the huge amount of data that flow in and out and how knowledge is sifted from this mass information is of critical importance to the success of an organization. These days, several organizations, big and small, have a division or unit devoted completely to the management of information. An efficient knowledge management system takes care of not only the accessible information but also offers capacity for continuous refinement of the process, resulting in increased benefit from the intellectual property that the organization holds.

8. **Access to Global Networks**
   A global network is any communication network that is accessible to individuals throughout the whole world or the global society [35]. Global access allows organizations to source for talent outside their country of origin. This brings together professionals and specialists from all over the world to work together on the project. Increased knowledge sharing and superior innovation occur as an organization's human capital contributes to their understanding of world and local markets as well as best business practices. If you set out to a supermarket and pick up a few stuff off the shelf from electronics and white goods or even clothing and look at the labels, the likelihood is that you will discover them having been produced in China or Mexico. The coffee pods you purchase to utilize for your daily use comes from Africa. Worldwide markets are growing outside borders and re-defining the way demand and supplies are managed. Global businesses across the world are driven by the marketplace.

9. **Case Studies**
   These are narratives that are used as a coaching tool to demonstrate the use of a theory or concept in real circumstances. Depending on the objective they are meant to accomplish, cases can be fact-driven and deductive where there is an acceptable answer, or they can be perspective-driven where many solutions are achievable.

   A good case study, according to Professor Paul Lawrence, is:

   > The vehicle by which a chunk of reality is brought into the classroom to be worked over by the class and the instructor. A good case keeps the class discussion grounded upon some of the stubborn facts that must be faced in real life situations. [36]

   Even though they have been used for the most part comprehensively in the teaching of medicine, law, and business, case studies can be an effectual training tool in any number of disciplines. As a teaching strategy, case studies have a number of qualities—"They bridge the gap between theory and practice and between the academy and the workplace" [37]. They also give students practice identifying the parameters of a problem, recognizing and articulating positions, evaluating courses of action, and arguing different points of view [37].

10. **Learning Seminars**
    A seminar is by and large understood to be a small group gathering in which students and tutors discuss information on a selected topic. They may perhaps be referred to as something else such as tutorial groups. Seminars present a chance to look at topics by discussion, and to recognize and sort out any problems. A number of tutors may use the opportunity to initiate new correlated topics. Seminars need not essentially be face-to-face contact; they can as well take place in online environments.

    Seminars present opportunities to:
    a. Explore topics in more depth
    b. Share ideas in a way that will advance your thinking
    c. Learn from other people's experiences and background knowledge
    d. Gain perspectives and points of view that you might not have otherwise considered
    e. Identify and sort out any misunderstandings

11. **Process-Oriented Practical Pilot**
    It is mainly concerned with "how" things are accomplished. It is an inclination to remain open and follow a new path. It refers to setting aside conventional ways of achieving results and instead following culturally respectful procedures that also generate results.

    Process-oriented practical pilot is demonstrated with the behaviors listed below [38]:
    a. Shows interest in new ways of doing business that work better for Aboriginal people
    b. Accepts that ways of being and doing differ across cultures
    c. Challenges self to follow Aboriginal leadership in determining and facilitating the process
    d. Demonstrates understanding focusing on the process of building relationships
    e. Adapts readily to a change in process
    f. Adapts to a range of different social and cultural situations
    g. Prepares for meetings and interactions by learning the appropriate protocol (from others in the BC Public Service and Aboriginal people), the intent and meaning behind the protocol, and when it should be used in a particular setting
    h. Shares early, openly, and honestly any time constraints that may influence the process
    i. Welcomes a meeting agenda that emerges from dialogue

j. Acknowledges that other ways of doing business are valid and valuable
k. Adapts business timelines and expectations to reflect flexibility
l. Plans contingencies for when the process may take longer than expected
m. Takes time to build relationship prior to doing business

## *Management perspective of engineering leadership*

> Managers Do Things Right; Leaders Do the Right Thing. [39]

The goal of any managing position is to inspire and empower people, to move a team from an idea to implementation, and finally to the completion of a project. Many of us are already managers in that our daily lives—school, work, and social demands all require some degree of management and organization. However, there is a substantive and discernable difference between management and leadership, and being able to lead by example:

> There Is a Difference Between Leadership and Management. Leadership Is of the Spirit, Compounded of Personality and Vision; Its Practice Is an Art. Management Is of the Mind, A Matter of Accurate Calculation—Its Practice Is a Science. Managers Are Necessary; Leaders Are Essential. [40]

So what do these quotes mean to an aspiring leader? Management is a role many feel comfortable in to varying degrees, and is not too far off, despite differences, from leadership. It is truly the quality of passion that is the key difference. How does management coincide with leadership? Consider the following quote:

> So for our careers, we must be managers as well as leaders. In other words, we must engage in the activities on a day-to-day basis (manager role) that will result in achieving the long-term vision we have for our career and life impact (leadership role). [41]

Any good manager will understand their team members and the technical skills each individual is proficient in. When matching these skills to the problems to be solved along with the team's objectives, you will be able to develop a management perspective on engineering leadership.

## *Fundamental differences between the role of a manager and a leader (Table 1.1)*

### *How leadership and management overlap*

> Management is efficiency in climbing the ladder of success; leadership determines whether the ladder is leaning against the right wall. [42]

### *Behavioral differences*

Zaleznik [43] upholds that managerial culture emphasizes rationality and control. Raised under this belief, managers are more inclined to be problem solvers by intuition, and their energies are spent on discovering solutions to the problems relating to managerial goals, resources, structures, and people [42,43]. This is why, contrary to leaders, managers are more systematic in nature, structured and purposeful in their approach, commanding and stabilizing in their behavior, and unrelenting and tough minded in their routine. A leadership culture, on the contrary, is open, forthcoming, candid, and participative. Consequently, it encourages the growth and assent of new ideas to tackle challenges. Taking the problems as opportunities, leaders seek new options and convince their followers to productively tackle the challenges.

Leaders are more rebellious in nature while managers prefer to conform to the organizational norms, rules, and hierarchy [44]. Therefore, most leaders challenge the status quo whereas managers prefer to accept the status quo [45]. In order to achieve better results, management strives to realize organizational efficiency along with effectiveness within the parameters of the organization's mission. However, leadership takes a different approach.

Perloff [46] argues that leadership creates and sells its visions to those who need to implement them, and evaluates whether these have been successful, along with determining what the next steps are. He uses an analogy of "trains" to describe the difference between leaders and managers. In his view, managers make the trains run on time, but it is the leaders who decide the destination as well as what freight and passengers the trains carry. Put simply, managers are more like tacticians, whereas leaders are strategists. Covey et al. [42] make the same point in a different way: management works within the established paradigm while leadership creates new paradigms. Management operates within the established system whereas leadership improves the existing systems and establishes more and better systems.

Leaders provide vision and inspiration, and support the people to do things, whereas managers provide the resources and expect results.

***Table 1.1*** Difference between leaders and managers

| No | Leaders | Managers | Reference |
|---|---|---|---|
| 1 | Leaders are change agents. | Managers are principally administrators. | |
| 2 | Leaders get organizations and people to change. | Managers write business plans, set budgets, and monitor progress. | [47] |
| 3 | Leaders select talent, motivate, coach, and build trust. | Managers plan, budget, evaluate, and facilitate. | |
| 4 | Leaders are more about soul or heart rather than mind. | Managers are more about mind. | |
| 5 | Leaders are visionaries, passionate, creative, flexible, inspiring, innovative, courageous, imaginative, experimental, and initiators of change. They draw their power from their personal traits and attributes. They make use of their referent power to influence their followers. | Managers are rational, consulting, persistent, problem-solving, tough minded, analytical, structured, deliberate, authoritative, and stabilizing. They draw their power from their position and authority. | [48] |
| 6 | Leaders have good intuition and insight. | Managers have good analytical ability. | |
| 7 | Leaders are inspiring visionaries concerned about substance. | Managers are planners who have concerns about the process. | [43] |
| 8 | Leaders set a direction, communicate it to everyone who will listen and probably many who won't, and keep people psyched when times get tough. | Managers establish systems, create rules and operating procedures, and put into place incentive programs and the like. | |
| 9 | Leaders are mobilized by their personal power and endorsement of the group. | Managers are mobilized by authority and position power. | |
| 10 | Leaders are strategists. | Managers are tacticians. | |
| 11 | Leaders leave a great deal to chance. | Managers are eager to solve the problems. | |
| 12 | All leaders are good managers. | All managers may not have leadership qualities. | [49] |

Zaleznik [43] suggests that leaders develop fresh approaches to long-standing problems and open issues to new options; managers act to limit choices. While leaders inspire the purpose, managers are concerned about systems, controls, procedures, policies, and structure [49]. The main role of the leaders is to set a new direction for a group. However, managers control, guarantee discipline, and introduce order according to established principles [50].

## *Technology leadership in engineering*

Technology leadership in engineering is about being a leader in innovating, that is, creating new ideas and solutions and new products that execute those new solutions. It is based on leading a team in the development of new ideas or solutions in the service of better and more effective products, processes, technologies, and ideas all of which are highly valued by markets, governments, and the wider society. This requires "open box" thinking, which hopefully is more likely and probable since you'll be closer to the problem than you otherwise would be at a higher management level.

Open box thinking (also thinking out of the box or thinking beyond the box) is a metaphor that means to think differently, unconventionally, or from a new perspective. This phrase often refers to novel or creative thinking. The term is thought to have been derived from management consultants in the 1970s and 1980s challenging their clients to solve the "nine dots" puzzle, whose solution requires some lateral thinking [51].

According to a common student reference, Wikipedia, the catchphrase, or cliché has become widely used in business environments, especially by management consultants and executive coaches, and has been referenced in a number of advertising slogans. To think outside the box is to look further and to try not thinking of the obvious things but to try thinking of the things beyond them [51]. Alternatively, some professionals have been told by leadership to go "beyond the box." In other words, they tell employees they don't want them thinking outside of the box, because they don't want them "inside the box" in the first place! However, to get outside of the box in the beginning, an organization must be solidly managed with a culture of creativity, innovativeness, and risk taking.

## *Seven steps for effective leadership*

A business is a gathering of people working toward a common goal, and a leader is needed to describe that cause. Business growth will only come in the course of the time and talents of others. To recognize, attract, fill, and maintain business leadership talent, companies need leadership improvement programs focused on employing strategies, employee improvement, and career and succession preparation.

Corporations face two major challenges in discovering and developing leaders. They need to identify capable candidates to fill present and future leadership roles, and they need to develop an all-inclusive leadership program to nurture and develop the leaders of tomorrow.

Now technology can be deployed to extend these practices across the enterprise and to all levels of the workforce. Human resource development then becomes an enterprise-wide effort across the management team and not just the job of the human resource department.

## *Elements of leadership development programs*

Key talent management roles all play a part in a broad leadership development program and can be well supported by an integrated talent management technology platform. These roles include employment, evaluation, performance management, succession planning, and career development.

A successful leadership development program begins with the alignment of leadership development with company strategy and an understanding of the types of leadership style(s) needed to execute that strategy. John Hansen [52] recommended seven steps for effective leadership development as listed below:

1. **Determine the Best Leadership Style for Your Organization**
   There are several assumptions and methods for determining the right leadership style for an organization. The leadership style, for example, that is required by a head of corporate security would clearly be vastly different from the leadership style of an art museum director. Company background will also play a major role in determining the leadership style. There are two ways to assess leaders' fit.
   a. Get to know them better. Psychological and behavioral assessments have been statistically linked to current and future success in leadership roles.
   b. Understand the culture better. Ask your board, employees, vendors, or consultants for insight into what makes an effective leader in the company.

      Use both sets of information to find alignments or disparities. If there is a glaring cultural conflict, be ready to find a better candidate who possesses the unique skills your organization requires.

2. **Identify Current and Potential Leaders Within or Outside the Company**
   Leaders can be found both within and on the outside. Companies must consider the cost and timing of rising internal leadership against the cost and availability of hiring from the outside. Study has shown that one of the key benefits of developing leaders on the

inside is that they achieve production almost 50 percent quicker than external candidates.

To assess potential leaders in the organization, a leadership program needs to recognize the expected management skills and competencies. Whether companies develop a competence model of their own or use an external model, they are required to classify the success measurements and incorporate them into their performance management system. When the leadership role cannot be filled from within the company, employment should use the same capacity to test the existing competencies of future prospective candidates.

Online pre-assessments and full evaluation testing can assist in ensuring that the right candidates are shortlisted. Untrained candidates are automatically filtered out, not on the basis of their resumes but, rather, on the basis of a self-administered online test or questionnaire.

3. **Identify Leadership Gaps**
   To fully recognize leadership gaps, companies must determine present and future leadership conditions and compare those with the existing leadership team. Then look at the leadership development pipeline and identify gaps in abilities and the time required to fill those gaps either via succession plan or recruitment.

4. **Develop Succession Plans for Critical Roles**
   Succession planning avoids disruption and employee trauma when the CEO leaves, whether the departure is anticipated or not. But a succession plan should not be confined to executive roles. As part of the leadership program, companies should evaluate critical roles throughout the organization. For the greatest efficacy, succession planning should be supported by technology systems that provide the ability to:
   a. Create backfill strategies that use data captured in the recruiting and performance review processes, coupled with individual career plans.
   b. Add multiple candidates to a succession shortlist and view all the best options.
   c. Display multiple talent profiles—from C-level executives to individual contributors—side by side to quickly identify the best fit.
   d. Track candidate readiness based on skills, competencies, and performance; promote top candidates based on relative ranking and composite feedback scores.

5. **Develop Career Planning Goals for Potential Leaders**
   Companies that support career planning for their employees gain in retention, engagement, and protection of the leadership pipeline.

Combining employee development with self-service career planning enables employees to investigate potential career paths and also select development activities necessary to attain them.

6. **Develop a Skills Roadmap for Future Leaders**
Once the high-potential workers have been identified, a skills roadmap should be developed for the future leaders. In today's connected world, growth programs need to support both conventional and nontraditional learning such as incorporating social networking tools into the development process.
7. **Develop Retention Programs for Current and Future Leaders**
Relating pay to performance can be a motivator for an employee, but goal position helps potential leaders stay focused on what is important to the company. Recognize excellent performance and base the upside of bonus potential on the success of both the employee and the company.
Leadership preservation is seriously imperative for all organizations, for two main reasons: turnover is costly and top performers drive best business performance.

### *Conclusion*

A well-designed leadership development program is the key to identifying, magnetizing, filling, and retaining corporate leadership. Technology applications can provide the enabling platform, including recruitment, assessments, performance management, succession and career planning, and development programs. Talent management practices implemented with robust technology applications can effectively identify and develop from all levels of the workforce leaders who will best drive business performance.

## *The lifelong learning imperative project: Findings*

The authors arrived at the following findings from their study and the conversations at the 2009 and 2011 workshops.

### *Reasons for change in U.S. graduate education*

A rudimentary lifelong learning infrastructure exists in the United States. Bourne et al. [53] note two predominant educational models: (1) at one extreme, continuing education programs blend with traditional degree-based programs. For example, courses developed for on-campus degree seekers are often slightly modified and repurposed as a short course offered to industry professionals; and (2) the other extreme, corporations contract with university faculty or for-profit vendors to develop continuing education content specific to their requirements.

## *Paradigm shift in the practice of engineering for innovation*

This rudimentary infrastructure is inadequate for today's (and tomorrow's) engineers. Evaluation in these courses is characterized by lack of standardization, and content is not uniform even within engineering sub disciplines. Very little is done to address the changing needs of learners, especially those who want to study and interact online. The most common approach to recognize formal, nondegree learning is by means of certificates.

## *Unlocking the innovative potential of the U.S. engineering workforce*

Stimulating lifelong learning in the United States will improve the knowledge base of the country's engineers and our capacity for innovation and competition. In particular, a national vision and actionable strategy to overcome barriers for lifelong learning in the engineering profession are required.

## *Fueling U.S. innovation by developing engineers in industry*

A well-coordinated effort between industry, academia, professional societies, and policymakers to develop a national framework for lifelong learning for engineers should begin as soon as possible. *The New York Times* reported that, based on a recent Battelle Memorial Institute study, Chinese spending on R&D will likely match U.S. spending in 2022 [54]. The article goes on to say that "if U.S. government labs, university departments and corporate researchers aren't already on top of the next generation of breakthroughs, the country will very likely fall behind in 10 or 20 years when those innovations become marketable products" [54]. Such a scenario is possible but not likely if American engineers, who are motivated to maintain and upgrade their skills, find it straightforward to access lifelong learning.

## *Conclusions*

The future of U.S. competitiveness and development can be improved by a robust lifelong learning system for engineers. Such a system would probably have numerous moving parts concerning universities, manufacturing, professional societies, and others and could be facilitated by a suitable government policy. New lifelong learning content and formation should take account of the needs of an assorted workforce (i.e., managerial and nonmanagerial, ethnic minorities and whites, men and women).

## *Intangibles of engineering leadership*

The *Intangibles of Leadership* uncovers models in the attributes that actually distinguish those who thrive at the top. After more than a decade of senior executive assessments, CEO interviews, and proprietary research, Davis found that extraordinary leaders possess certain characteristics that fall between the lines of existing leadership models, yet are fundamental to executive success. Davis explains each of these qualities, the people who exemplify them, how to detect them in others, and how to develop the subtle characteristics that will enable leaders to stand out from the pack. The book has been highly reviewed and was named as a "Top Business Book of 2010 by Library Journal" [55].

According to Davis, here are the 10 defining characteristics that make for extraordinary leaders [56]:

1. Wisdom based on experience, reflection, and perspective. You're advice-worthy. You exercise good judgment. You think independently and don't speak in banalities.
2. The will to stand firm. You never give up and you make your own luck. "Successful leaders I've encountered over the years enable things to happen, rather than wait for them to happen," says Davis.
3. Executive maturity to read and understand how others are feeling and also to control your own emotions. You stay cool under pressure and know how to navigate your way through a crisis.
4. Integrity built on trust, consistency, and a moral compass. You don't lie. Don't cheat. And you always keep your promises.
5. Social judgment or the ability to analyze people and situations and then make smart decisions. In other words, excellent leaders have a consistent ability to connect with people.
6. A presidential presence, drawing on reputation, identity, charisma, and superior communication skills.
7. Self-insight, a key trait that often decides whether Davis recommends a candidate for the executive suite. It's about knowing your strengths and weaknesses, understanding your hot buttons and blind spots, and recognizing your impact on others. Figure out what makes you tick. It's the only way you'll get better at what you do.
8. Self-efficacy, with a deep faith and fundamental belief in your ability to get a specific job done. Not only do you want the ball. You also know you're going to run it into the end zone.

9. Fortitude, courage, and resilience. Simply stated, people with fortitude are thick-skinned and develop an ability to not let things bother them to the point that it impacts progress toward a goal.
10. Fallibility, with a willingness to show rather than hide your flaws. Acknowledge and embrace your imperfections. Fess up that you don't have all the answers. In fact, it can be said that the most outstanding leaders are successful because they are fallible, rather than in spite of it.

According to the author, he wrote the book as designed to provide insight into the personality traits that most describe extraordinary leaders and their underlying emotional mechanisms. It was premeditated as a mirror to help executives think about how they approach leadership and gain insights that can enable them to grow and develop.

## *Changing landscape for engineering leaders*

The requirements for and expectations of leaders in the engineering field are changing and evolving to keep up with the world around us. The three noteworthy primary drivers of these patterns of change are as follows:

- Enhanced pace of technological progress (compression of the innovation cycle)
- Globalization
- Dynamic economic climate

## *The pace of technological progress*

Innovation, the need for it and its execution, is the key driver of change in all aspects of life. Today, we see that the product development cycle has been greatly compressed and continues to be reduced. People who work in manufacturing and production don't just create products; they create them as quickly as possible, as inexpensively as possible, and in the necessary quantities [57]. With this we can see that more decisions must be made in shorter periods of time while more people are involved in decision-making which risks slowing the process, and the overall development cycle continues to compress to ever-shorter turnaround timeframes.

The U.S. Architecture, Engineering, and Construction (AEC) industry is faced with the ever-increasing challenge of managing the public and private facilities and infrastructure to support the accomplishments of its economy. The increasing global emphasis on sustainable approaches and the need to increase efficiency and improve cost over the lifecycle of projects demand new approaches to AEC education. This study was initiated to look for insights into the current educational environment and to provide a baseline for possible solutions to cope with the complexity of the challenge.

The study, carried out by Becerik-Gerber et al. [58], examined 101 U.S. AEC programs focusing on emerging subject areas of Building Information Modeling (BIM) and sustainability, and reviewed how educational innovations of distance learning, multidisciplinary collaboration, and industry collaborations are incorporated to develop core competencies in those two subject areas. The researchers reviewed and categorized the AEC disciplines based on the respective accrediting bodies of ABET, NAAB, and ACCE, and surveyed the internal factors (program resources, expertise, etc.) and external factors (accreditation requirements, sustainability initiatives, etc.) that affect the pedagogical approaches.

The study illustrates the challenges in incorporating new knowledge areas into constrained curricula and the various approaches that the university programs are undertaking. A comparative analysis also reveals the similarities and differences and specific advantages and disadvantages of particular approaches across the AEC programs. The findings reinforce the notion that there are disparities in these educational programs, which need realignment to develop the workforce of the future that will lead the AEC industry transformations.

## *Globalization*

The fact that we are in a global society has an additional significant implication for the outcomes of engineering leadership. Specifically, the results of our technological and leadership efforts have the potential to impact the lives of individuals across continents and around the world. Thus, the imperative to focus on quality engineering leadership is further amplified. The globalization of the economy and society means that people halfway around the world are talking about the same topics and addressing similar challenges simultaneously. With multinational teams and broader customer bases comes the blessing and challenge of higher levels of competition. Engineering leadership needs to take a global perspective to be able to consider the impact on global customers. This also must apply to engineering teams to be able to take into account the different cultural expectations and local social norms among team members.

Leadership in science and technology gives the United States its comparative advantage in the global economy. U.S. exports are majorly from sectors that rely extensively on scientific and engineering workers and that embody the newest technologies. In 2003, with a massive national trade deficit, the smallest deficit relative to output was in high-technology industries. Aggregate measures of scientific and technological prowess place the United States at the top of global rankings.

Scientific and technological preeminence is also critical to the nation's defense, as evidenced by the employment of R&D scientists and engineers in defense-related activities and in the technological dominance of the U.S. military on battlefields. To be sure, other factors also contribute to U.S. economic leadership, but in a knowledge-based economy, leadership in science and technology contributes substantially to economic success [4].

The paper presents evidence that changes in the global job market for S&E workers is eroding U.S. dominance in S&E and that the erosion will continue into the foreseeable future, diminishing the country's comparative advantage in high-tech goods and services and threatening the country's global economic leadership. The paper assesses policies that could smoothen the transition of the United States from being the superpower in S&E to being one of many centers of excellence [5].

The analysis can be summarized in four propositions, two relating to the job market for scientific and engineering talent, and two relating to the effects of that market on the economy. The propositions regarding the S&E job market are: The U.S. share of the world's S&E graduates at all degree levels is declining rapidly, as college enrollments have expanded in other countries. The number of S&E PhDs from European and Asian universities, particularly from China, has increased while the number from U.S. universities has stagnated. International students have, in addition, increased their share of advanced S&E degrees from U.S. universities. As a result, U.S. reliance on foreign-born scientists and engineers has increased [5].

The job market for young scientists and engineers in the United States has worsened relative to job markets for young workers in many other high-level occupations, which discourages U.S. students from venturing into these fields. At the same time, rewards are sufficient to attract large immigrant flows, particularly from less developed countries. The propositions regarding the impact of changes in the supply of S&E talent on the country's economic performance are:

> By increasing the number of scientists and engineers, highly populous low-income countries such as China and India can compete with the United States in technically advanced industries even though S&E workers are a small proportion of their workforces. This threatens to undo the traditional "North–South" pattern of trade in which advanced countries dominate high-tech manufacturing while developing countries specialize in less-skilled manufacturing. [4]
>
> Diminished comparative advantage in high-tech manufacturing will create adjustment problems for U.S. workers, of which the offshoring of IT jobs to India, growth of high-tech production

> and exports from China, and multinational movement of R&D facilities to developing countries are harbingers. The country faces a long transition to a less dominant position in science- and engineering-associated industries, for which the United States will have to develop a new labor market and R&D policies that build on existing strengths and develop new ways of benefiting from scientific and technological advances in other countries. [4]

According to Professors Pisano and Shih, these U.S. companies should re-evaluate the approach to manufacturing operations. The following summarizes their research-based assessment of U.S. companies outsourcing manufacturing [4]:

---

For decades, U.S. companies have been outsourcing manufacturing in the belief that it held no competitive advantage. That's been a disaster, maintain Harvard professors Pisano and Shih, because today's low-value manufacturing operations hold the seeds of tomorrow's innovative new products.

What those companies have been ceding is the country's *industrial commons,* that is, the collective operational capabilities that underpin new product and process development in the U.S. industrial sector. As a result, America has lost not only the ability to develop and manufacture high-tech products like televisions, memory chips, and laptops but also the expertise to produce emerging hot products like the Kindle e-reader, high-end servers, solar panels, and the batteries that will power the next generation of automobiles.

To rebuild the commons and restore its wealth-generating machine, the government and industry in the United States will have to make two drastic changes:

1. The government must change the way it supports basic and applied scientific research to promote the broad collaboration with business and academia needed to tackle society's major problems.
2. Corporate management practices and governance structures must be overhauled so that they no longer exaggerate the payoffs and discount the dangers of outsourcing production and cutting investments in R&D.

Restoring the ability of enterprises to develop and manufacture high-tech products in America is the only way the country can hope to pay down its enormous deficits and raise its citizens' standard of living.

---

## *The dynamic economic climate*

Today's global, interconnected economy means it is even more important that as engineers we're innovative about the creation and delivery of products because it impacts the bottom line. If your industry isn't moving forward, you're being left behind and you have no other choice but to have a progressive company environment with an innovate-first attitude that has been the hallmark of successful entrepreneurs for generations. For every problem, there is a better mousetrap!

The global economic crisis started in summer 2007, although the full impact was not felt till the bankruptcy of the investment bank, Lehmann Brothers in September 2008. The next couple of years witnessed heavy job losses and contraction in the gross domestic product (GDP) of many countries in the West as well as in the developing world. What started off with the subprime mortgage crisis quickly morphed into a full-fledged crisis of historic proportions prompting many commentators to draw parallels with the Great Depression of the 1930s.

## *Dynamic economic period of the great recession of 2008 and how it impacted businesses, innovation, and engineering globally*

The global economic crisis was caused by the coming together of several structural as well as business cycle factors that conspired to produce a perfect storm of epic proportions. These factors ranged from the collapse of the housing market in the United States, imbalances between the West and the East in terms of trade deficits, reckless and risky speculation, and finally the sovereign debt crisis that was a culmination of years of fiscal profligacy and loose monetary policies. The point about the global economic crisis, or the Great Recession as it is also called, is that the crisis exposed the chinks in the armor of the global economy and highlighted the pitfalls of too much integration and interconnectedness. Nowhere was this more apparent than in the aftermath of the collapse of Lehmann Brothers when the entire credit system froze and the global financial system came perilously close to collapse.

The global economic crisis basically originated in the West but had its effects on all economies of the world. Of course, the United States and Europe were the primary victims of the crisis and it can be said that countries like India and China were relatively unscathed in the wake of the crisis. However, this is not to say that these countries have successfully "decoupled" from the West since the tightly knit global economy and the dependence of China on exports to the United States for goods and India for services means that these countries have a fair amount of work to do before they can be called safe. The point here is that the United States and Europe were badly bruised by the crisis and it is still not clear when these countries and their economies would be out of the woods, if at all they would.

Finally, the global economic crisis has undone the many gains of globalization and hence there are renewed calls for protectionism and for erecting trade barriers in the West as well as in the East. This means that the global economic crisis has dealt a body blow to the global economy, which might take years to regain its earlier prosperity.

## *Where do engineers need to lead?*

The meticulous but realistic outlook of an engineering-educated mind is beneficial in multiple environments beyond traditional engineering problem-solving. The problem-solving, analytical, and fact-based approaches used in engineering solutions are key strengths for an effective leader. The tech-savvy, understand-how-it-all-works mindset can introduce a competitive advantage to organizations and nations in the global interconnected economy.

At some point in the past 30 years, the perception of what an engineer could do shifted. As a result, most individuals with PhDs moved on to traditional careers such as researchers at some of the national labs or into professorship. Historically, however, some of our greatest business and public leaders have hailed from the engineering field: John Frank Stevens managed the construction of the Panama Canal; among others were Thomas Edison, Henry Ford, Herbert Hoover, and Jimmy Carter. At some point between then and now, things shifted. The American shift away from a primarily industrial economy to the present day's more service-based economy meant engineers were gradually replaced by professional business managers resulting in the loss of the leadership mentality among engineers with a corresponding loss of an engineering mentality across the wider business community.

We're going back to our roots, but in the best of ways. Today, 20 percent of Fortune 500 CEOs have engineering degrees and opportunity is blossoming further. Today there is many different routes to be taken, as engineers explore diverse disciplines such as academia, private industries that include large multinational companies, national corporations, small businesses and entrepreneurial endeavors. It is truly inspiring to see more and more PhDs taking up the mantle of entrepreneur and starting their own businesses.

The future continued success of U.S. businesses and industry rests upon having more practicing engineers with knowledge in topics such as: principles of leadership and management, global/international impact, ethical standards, cultural diversity, conflict resolution, and communication skills [59]. Therefore, it is critical for U.S. competitiveness that future engineering graduates possess strong technical engineering skills as well as other developmental skills such as leadership and management.

To be effective leaders, engineers must possess skills such as written and oral communication, customer relations, personal initiative, teamwork abilities, organizational knowledge, and decision-making that will facilitate the development of solutions to business challenges [60].

Engineering leadership is the ability to lead a group of engineers and technical personnel responsible for creating, designing, developing, implementing, and evaluating products, systems, or services. To appropriately prepare engineering leaders of the future, it is imperative to be aware of the necessary skills required to successfully complete engineering leadership roles, responsibilities, and positions.

To further understand the skills needed for successful engineering leadership development, a research study was conducted with engineering students and professionals. The study consisted of developing and distributing surveys to obtain feedback from engineering students and professionals on key topics related to engineering leadership

## *Engineering leadership development programs: A look at what is needed and what is being done—Research finding*

The research findings presented in the above sections of this paper focus on presenting feedback from both engineering professionals as well as current engineering students. Surveys were distributed at different conferences, resulting in responses from more than 500 participants [61].

Significant results obtained include:

- Professional engineers indicated that they are most confident in their ability to solve problems, lead a team, and listen to others. They were least confident in their ability to give oral presentations, write technical papers, and be persuasive.
- Students considered themselves proficient at listening to others and thinking critically but rated themselves lowest on writing technical papers and public speaking. Overall, communication and problem-solving skills were considered to be the most useful when performing in a leadership position, while

counseling and controlling group performance were identified as the least useful.

- Communication skills were identified by both students and professional participants as one of the most important skills for engineering leaders.
- The most important areas to prepare for leadership roles were identified as knowing where to fit within the organization, mentoring, people skills, negotiation skills, understanding team limits, time management, communication skills, resource leverage, being open-minded, the ability to develop a vision, and being a good listener.

Survey results reveal that the most beneficial areas of leadership training were people skills dealing with different people and personalities, team dynamics, ethics, project management, cross-functional projects, globalization, planning, facilitation and communication skills, strength discovery, conflict resolution, cross-cultural communication, and learning from mistakes.

## *Conclusion*

Due to the competitive nature of businesses, industry is demanding not only technically proficient engineers for their companies but also engineers that are prepared to take on leadership positions. To be effective leaders, engineers must possess the soft skills necessary to solve business challenges. These skills include written and oral communication, self-initiative, teamwork abilities, customer relations, and decision-making.

Engineers and future engineering graduates are needed to assume leadership positions, from which they can serve as positive influences in the making of public policy and in the administration of government and industry. Current engineering leadership programs, along with those currently under development, will need to include additional topics. The need to develop a holistic engineering leadership program that entails the aforementioned skills such as the ability to control a group, think critically, be a visionary, inspire, influence, adapt, maintain good interpersonal relations, communicate well; and also traits like being open-minded, people-centered, action-oriented, equitable, likeable, determined, confident, credible, honorable, fair, and a networker have been identified within this research as capabilities and traits essential to ensure that engineering professionals and future engineers are prepared to flourish as leaders.

**Case study: Recognizing leadership opportunities**

Whether you're an intern or the CEO, you have the opportunity to be a leader. Each rung of the leadership ladder offers different opportunities, and you need only to grasp the nearest leadership opportunity within reach to pull yourself up.

That's why in order to groom your leadership skills, you must first learn to recognize leadership opportunities. Leadership opportunities take many forms. Sometimes it means taking on a project that nobody wants to do and rocking it! Or perhaps you find a better way to accomplish an everyday task within your organization, and recognize this as an opportunity to raise your hand and bring in a new or different perspective.

We often hear the expression "lead by example." This is perhaps the easiest leadership opportunity there is, and leaders who work hard and are proud of what they do tend to attract employees who want to follow that lead. Leading by example can be seen in the way you dress, interact with others, and carry yourself. Impressions matter, so do your best to project a confident, positive presence wherever you go, to whomever you encounter.

Another great way to find leadership opportunities is to ask your boss or manager. Maybe there's a project that has been sitting on your boss's desk for months and he or she needs someone to pick it up and run with it. Be proactive about letting your organization's leaders know that you are ready to take on more responsibility and would like an opportunity to tackle new challenges. Even if there aren't any projects available at that moment, your boss will keep you in mind next time a project rolls around that needs a leader, and will appreciate that you took the initiative to step up to the plate.

There are also plenty of leadership opportunities to be found outside your place of work. Brush up on your leadership skills by volunteering or joining a youth mentorship program like Big Brothers Big Sisters of America. Being a role model for a child in need is a fantastic way to polish your leadership skills while making a positive difference—something every leader strives to do! You can also join clubs or interest groups at your school to gain leadership skills. Getting involved in a group that you're passionate about will always make you a better leader, so you shouldn't feel obligated to join a club or take a volunteer position that doesn't fit your interests. Can't find a club that suits your interests? Start your own! If you want to gain meaningful leadership skills you need to practice, so seek out a leadership opportunity and start putting those skills to use.

**QUESTIONS**

1. You just started your first day at a new company. How do you go about finding leadership opportunities for yourself?
2. Explain how you might use one of the tips mentioned above to position yourself as a leader.

---

### Real profiles in engineering leadership

**Name:** Stephanie C. Hill

**Current position or field of expertise:** President, Lockheed Martin Information Systems & Global Solutions—Civil

**What I like most about my position:** In my role, I have the opportunity to work with amazing people on important missions for customers that can make a huge difference in the lives of citizens both in the United States as well as several other countries. There are not too many positions where you know what you're doing is noble work. I have the privilege of serving all the civilian, nondefense parts of the U.S. government, commercial customers, and international governments. Our wonderful team consists of about 10,000 employees in all 50 states and in 9 countries. We leverage technology to support our customers' missions from providing air traffic management systems that control 60 percent of the world's air space to delivering energy efficiency programs and cyber security to utilities and federal customers—just to name a few.

**Who or what has made a difference in my career:** My parents have been the most influential people in my career. They gave me such a strong balance of committing myself to excellence and genuinely caring about the people around me. I learned this by watching them because they cared about the people around them and their family and they showed a commitment to excellence and effective leadership in their actions every day.

When I would come home with a 95 score, my daddy would ask me "Where are the other 5 points?" He taught me that you have to work hard so that you don't leave those 5 points on the table. But my mom balanced that approach by being supportive and celebrating my 95. If I hadn't had these two examples, my leadership style would've been much different.

Another big influencer was my grandmother. She worked full time as a maid until she was 93. Her work ethic was second to none. And I never heard her say a negative thing about anyone.

**I felt like quitting when:** My family instilled in me a solid work ethic where you give 120 percent to any endeavor and quitting is never really an option. While I have had a stimulating and rewarding career, there certainly have been challenges and frustrations along the way. But I faced those head on—driven by my commitment to my teams, to our customers, and to Lockheed Martin. I've been with the company for 26 years and can honestly say I have never been bored, and have always felt grateful for the opportunities and experiences gained during my tenure.

**My strategies for success are:** I live and work by these four basic tenets: trust, transparency, engagement, and collaboration. When these are all present in any aspect of life, the sky is the limit for how far you can go.

In addition, these are some growth principles I share with others on how to be successful in career development:

- Create a track record of success
- Choose the right attitude
- Be authentic
- Build business acumen
- Step outside your comfort zone
- Engage and care about the team around you

**I am excited to be working on:** My business is responsible for supporting federal government customers and their critical missions. Our teams ensure the nation's security by providing next generation biometrics to the FBI, guarantee global air traffic safety by supporting the FAA, explore space and support scientific research through our contracts with NASA and the NSF, help Americans live healthier lives as we support claims processing and disability examinations for millions of veterans, and address energy challenges by implementing energy efficiency programs. The portfolio is not only diverse but also impactful so that there is no shortage of excitement!

My other passion is promoting science, technology, engineering, and mathematics (STEM) education. It is a national imperative to increase the number of U.S. students, especially minorities and women, in STEM fields. Along my career journey, I have had mentors push me, guide me, and ultimately, inspire me. And I want to do the same. Students need to know about the diverse—and exciting—opportunities available in STEM fields. The student inspired today could be the one to help solve our nation's critical challenges tomorrow. We have to encourage our young people to know that anything is possible for them!

**My life is:** My life is fulfilling both professionally and personally. I am blessed and truly grateful to have an amazing family. And on top of that, my role at Lockheed Martin allows me to contribute and make a difference for our workforce, our customers—and their critical missions—and our community.

*Source:* Transforming Your STEM Career, Inc. Available at: TransformingYourSTEMCareer.com website

## References

1. Musselman, C. (2010). Leadership in engineering: Why is that important in engineering education? Available at: http://www.nspe.org/resources/blogs/pe-licensing-blog/leadership-engineering-why-important-engineering-education#sthash.kpSomUHx.dpuf
2. Veblen, T. (1921). *The engineers and the price system*. Kitchener: Batoche Books.
3. Freeman, R.B. (2006). Does globalization of the scientific/engineering workforce threaten U.S. economic leadership? In *Innovation Policy and the Economy, volume 6,* Adam B. Jaffe, Josh Lerner and Scott Stern (eds.), The MIT Press, p. 124.
4. Shanghai JiaoTong University's Institute of Higher Education. A substantial proportion of who are immigrants: Physics 32 percent; Physio/Med 31 percent; Economics 31 percent; Chemistry 26 percent. Available at: http://ed.sjtu.edu.cn/rank/2004/2004Main.htm
5. National Science Foundation. (2014). National Center for Science and Engineering Statistics (NCSES) Science and Engineering Indicators 2014, Arlington, VA (NSB 14-01).
6. National Science Board. Science and engineering indicators 2014 digest. Available at: http://www.nsf.gov/statistics/seind14/index.cfm/digest
7. Galama, T. and Hosek, J. (2008). *U.S. Competitiveness in Science and Technology,* MG-674-OSD, pp. 188.

8. Gordon, B.M. MIT Engineering Leadership Program. Available at: https://gelp.mit.edu/student, assessed on October 12, 2016.
9. Institute for Leadership Education in Engineering: What is engineering leadership education? Available at: http://ilead.engineering.utoronto.ca/about-ilead/what-is-engineering-leadership-education/, accessed on October 12, 2016.
10. The STEM Workforce Challenge: the Role of the Public Workforce System in a National Solution for a Competitive Science, Technology, Engineering, and Mathematics (STEM) Workforce. (2007). Report prepared for the U.S. Department of Labor, Employment and Training Administration by Jobs for the Future. Available at: https://www.doleta.gov/youth_services/pdf/STEM_Report_4%2007.pdf, accessed on October 12, 2016.
11. Arnold, G. From Dilbert to Lumberg: the evolution of an engineering leader. Available at: https://blog.linkedin.com/2011/10/05/engineering-leader-evolution, accessed on November 11, 2016.
12. TechCrunch. Dropbox acquires Zulip. Available at: http://techcrunch.com/2014/03/17/dropbox-acquires-zulip-a-stealthy-workplace-chat-solution-still-in-private-beta/
13. McKellar, J. *This Is What Impactful Engineering Leadership Looks Like*. Available at: http://firstround.com/review/this-is-what-impactfulengineering-leadership-looks-like/, accessed on October 12, 2016.
14. The International Labor Organization (ILO). *Reuters* (2008-10-21). Financial crisis to cost 20 mn jobs: UN. Available at: http://www.expressindia.com/latest-news/Financial-crisis-to-cost-20-mn-jobs-UN/376061/
15. U.S. Department of Labor, Bureau of Labor Statistics. (2008). The employment situation: January 2008. Available at: http://www.bls.gov/news.release/pdf/empsit.pdf
16. Goodman, P.S. (2009). U.S. unemployment rate hits 10.2%, highest in 26 years. *The New York Times*. Available at: http://www.nytimes.com/2009/11/07/business/economy/07jobs.html
17. Sheppard, S.D. and W. Sullivan. (2008). *Educating engineers: Theory, practice, and imagination*. Palo Alto, CA: Carnegie Foundation for the Advancement of Teaching.
18. The National Innovation Initiative. (2003). *Innovate America: Thriving in a world of challenge and change*. Washington, DC: Council on Competitiveness.
19. National Academy of Engineering. (2004). *The engineer of 2020: Visions of engineering in the new century*. Washington, DC: National Press. Available at: http://www.nap.edu/catalog/10999.html
20. Fasano, A. (2014). *Engineer your own success: 7 key elements to creating an extraordinary engineering career, updated and expanded*. Wiley-IEEE Press, Hoboken, NJ, 232 p.
21. All Engineering Schools. Available at: http://www.allengineeringschools.com/engineering-degree/all-degrees/engineering-management/california
22. Tufts University. MSEM vs. MBA. Available at: http://gordon.tufts.edu/programs/m-s-in-engineering-management/msem-vs-mba
23. Meadows, D., et al. (2004). *Limits to growth: The 30-year update*. Chelsea Green Publishing. Available at: http://1a0.c26.mwp.accessdomain.com/archives/tools-for-the-transition-to-sustainability/
24. Ontario Building Envelope Council. A promise to future generations. Available at: https://obec.on.ca/sites/default/uploads/files/Promise-Form.pdf, accessed on October 12, 2016.

25. Chu, J., et al. (2010). Uncovering Leadership Opportunities—Realizing Your Potential as an Undergraduate. Engineering Leadership Review. Available at: https://obec.on.ca/sites/default/uploads/files/Promise-Form.pdf
26. Publishers Weekly From the archives: John F. Kennedy, 1917–1963. Available at: http://www.publishersweekly.com/pw/by-topic/industry-news/publisher-news/article/58542-from-the-archives-john-f-kennedy-1917-1963.html
27. *The Holy Bible*, Proverbs 29:18, the King James Version translation.
28. Griffiths, S., K. Houston, and A. Lazenbatt. (1995). *Enhancing student learning through peer tutoring in higher education*. Coleraine: Educational Development Unit, University of Ulste.
29. Boud, D.J. (1988). Moving towards autonomy. In: D. Boud (Ed.), *Developing student autonomy in learning* (2nd ed.). London: Kogan Page, pp. 17–39.
30. Carsrud, A.L., et al. (1984). Undergraduate psychology research conferences: Goals, policies, and procedures. *Teaching of Psychology*, 11, 141–145.
31. Twinning.org. A quick overview. Available at: http://www.twinning.org/en/page/a-quick-overview#.VsK0cEAd49o
32. Renton, J. (2009). *Coaching and mentoring: What they are and how to make the most of them*. New York: Bloomberg Press.
33. Wikipedia. Coaching. Available at: https://en.wikipedia.org/wiki/Coaching#cite_note-15
34. Stern, L.R. (2004). Executive coaching: A working definition. *Consulting Psychology Journal: Practice and Research*, 56(3), 154–162.
35. Wikipedia. Telecommunications network. Available at: https://en.wikipedia.org/wiki/Communication_network
36. Christensen, C.R. (1981). *Teaching by the case method*. Boston, MA: Harvard Business School.
37. Barkley, et al. (2005). *Collaborative learning techniques: A handbook for college faculty*. San-Francisco, CA: Jossey-Bass.
38. British Columbia Government. Human Resources for the B.C. Public Service. Available at: http://www2.gov.bc.ca/myhr/article.page?ContentID=4f3b094d-ff23-5e8b-fc22-1bfb6d226e80& PageNumber=2
39. Hawley, R. (2001). *Were you born to lead?* London: The Institution of Engineering and Technology, pp. 247–248.
40. Field Marshal Lord Slim. (2011). *Leadership quotes*. Available at: http://www.motivateus.com/leadership-quotes-02-11.htm
41. Bush, P.M. (2012). *Transforming your STEM career through leadership and innovation: Inspiration*. Academic Press: Elsevier, MA 182 p.
42. Covey, et al. (1994). *First things first: To live, to love, to learn, to leave a legacy*. New York: Simon and Schuster.
43. Zaleznik, A. (1977). Managers and leaders: Are they different? *Harvard Business Review*, 55(3), 67–78.
44. Kumle, J., et al. (1999). Leadership versus management. *Supervision*, 61, 4.
45. Bennis, W., and Nanus, B. (1985). *Leaders: The strategies for taking charge*. New York: Harper and Row.
46. Perloff, R. (2004). Managing and leading: The universal importance of, and differentiation between, two essential functions. Talk presented at Oxford University, Oxford, July 14–15.
47. Maccoby, M. (2000). Understanding the difference between management and leadership. *Research Technology Management*, 43(1), 57–59.

48. Capowski, G. (1994). Anatomy of a leader: Where are the leaders of tomorrow? *Management Review*, 833, 10–14.
49. Daft, R.L. (2003). *Management* (6th ed.). London: Dryden.
50. Schumpeter, J. (1934). *Capitalism, socialism, and democracy*, Vol. 14. New York: Harper and Row.
51. Wikipedia. Thinking outside the box. Available at: https://en.wikipedia.org/wiki/Thinking_outside_the_box
52. Hansen, J. (2012). Seven steps for effective leadership development. An Oracle White Paper.
53. Bourne, J., D.A.Harris, and F. Mayadas. (2005). Online engineering education: Learning anywhere, anytime. *Journal of Engineering Education*, 94(1), 131–146.
54. Davidson, A. (2011). Will China outsmart the US? *NY Times*, December 28. Available at: http://www.nytimes.com/2012/01/01/magazine/adam-davidson-china-threat.html
55. Library Journal. Available at: http://www.libraryjournal.com/lj/collection developmentspecialty2/888437-483/lj_best_business_books_2010.html.csp#management
56. Davis, B. (2009). What's a Global Recession? *The Wall Street Journal*. Available at: http://blogs.wsj.com/economics/2009/04/22/whats-a-global-recession/
57. Wetfeet. Career overview: Manufacturing and production. Available at: https://www.wetfeet.com/articles/career-overview-manufacturing-and-production
58. Becerik-Gerber, B., et al. (2011). The pace of technological innovation in architecture, engineering, and construction education: Integrating recent trends into the curricula. *ITcon*, 16, 411–432.
59. National Academy of Engineering. 2004. Available at: http://www.nap.edu/read/10999/chapter/1
60. Newport, C.L., et al. (1997). Effective engineers. *International Journal of Engineering Education*, 13(5), 325–332.
61. Crumpton-Young, L., et al. (2010). Engineering leadership development programs. A look at what is needed and what is being done. *Journal of STEM Education Innovations and Research*, 11(3), 10–21.

# *chapter two*

# *The role of creativity and innovation in leadership*

Innovation is widely heralded as essential for successful competition in the increasingly global economy. However, to enhance innovation in education, organizations and countries require transformative thinking. National thought leaders and organizations such as the National Academy of Engineering are supporting projects to explore this relationship. The Educate to Innovate (ETI) project was designed to explore the issue regarding teaching innovation and the expected outcome, entrepreneurship [1].

During the 1950s and 1960s, *Sputnik* and the space race stimulated a generation of Americans to follow education and careers in science and technology. Half a century later, American students are now graded 22nd and 21st among their peers all over the world in science and math, respectively. Students in the United States, formerly a leader in science, technology, engineering, and mathematics (STEM), are now outperformed by students from Slovenia, Hungary, and Estonia, among others [2].

In 1983, the National Commission on Excellence in Education published "A Nation at Risk," a nationwide study that highlighted the intolerable state of the American education system:

> Our nation is at risk. Our once unchallenged preeminence in commerce, industry, science, and technological innovation is being overtaken by competitors throughout the world. This report is concerned with only one of the many causes and dimensions of the problem, but it is the one that undergirds American prosperity, security, and civility. What was unimaginable a generation ago has begun to occur—others are matching and surpassing our educational attainments. If an unfriendly foreign power had attempted to impose on America the mediocre educational performance that exists today, we might well have viewed it as an act of war. [3]

More than two decades afterward, in 2010, the National Academies of Science, Engineering, and Medicine published *Rising above the Gathering Storm, Revisited: Rapidly Approaching Category 5*, which built on the findings of its 2005 "Gathering Storm" report. Notably, the report warns that, "Today, for the first time in history, America's younger generation is less well-educated than its parents" [4].

In an effort to respond to the faltering academic status of American students and in a quest to elevate them "from the middle to the top of the pack in science and math," the Obama Administration announced its ETI initiative in November 2009 [5].

President Barack Obama's ETI campaign is publicized as a joint effort between the federal government, the private sector, and the nonprofit and research communities to raise the standing of American students in science and math through dedication of time and money, and volunteering. The program attempts to enhance STEM literacy, improve teaching quality, and develop educational and career opportunities for America's youth.

At the time the program was first declared in November 2009, the participating organizations offered a financial and in-kind commitment of more than $260 million. Taxpayer commitments for the federal government's portion of ETI add to that total.

In addition, five public–private partnerships were announced, as well as commitments by key societal and private sector leaders to muster funds for STEM education, innovation, and awareness [6]. These partnerships and commitments are:

- Time Warner Cable's "Connect a Million Minds" (CAMM), which pledges to connect children to after-school STEM programs and activities in their area.
- Discovery Communications' "Be the Future" will broadcast dedicated science programming to more than 99 million homes and offer interactive science education to approximately 60,000 schools.
- *Sesame Street's* "Early STEM Literacy" commits to a two-year focus on STEM subjects.
- National Lab Day will promote hands-on learning with 100,000 teachers and 10 million students over the next four years and foster communities of collaboration between volunteers, students, and educators in STEM education. These initiatives will then culminate in a nationally recognized day centered on science activities.
- The National STEM Video Game Challenge promotes the design and creation of STEM-related video games.
- The annual White House Science Fair will bring the winners of science fairs from across the nation to the White House to showcase their STEM creations and innovation.

- Sally Ride, the first female astronaut, Craig Barrett, the former Intel chairman, Ursula Burns, CEO of XEROX, and Glenn Britt, CEO of Eastman Kodak, committed to foster interest and support for STEM education among American corporations and philanthropists [6].

In January 2010, President Obama announced the continuation of the program, stressing the half-billion-dollar monetary obligation from the administration's partners. This development includes an additional commitment of $250 million in financial and in-kind support, and a pledge by 75 of the nation's biggest public universities to train 10,000 new teachers by 2015. The program expansion also incorporated additional public–private partnerships anticipated to aid the training of new STEM educators, together with the launch of Intel's Science and Math Teachers Initiative and the PBS Innovative Educators Challenge, as well as the expansion of the National Math and Science Initiative's UTeach program and Woodrow Wilson Teaching Fellowships in math and science. In addition, the president called on 200,000 federal government staff working in the fields of S&E to volunteer to work with educators in order to foster enhanced STEM education [6].

STEM-educated workforce is very important for the protection and the wealth of the United States as industry and government increasingly demand exceedingly trained STEM professionals to vie in the international market, and look to science and technology to help stay one step ahead of national security threats.

The United States must not permit itself to be outcompeted in science, technology, engineering, and mathematics. While the administration's ETI enterprise is projected to raise the United States "from the middle to the top of the pack in science and math," this one-size-fits-all federal approach fails to cure the primary problems of educational performance and does not stop the permeable pipeline in the American education system.

## *The evolution of innovation*

The principles associated with innovation can be applied to organizations, individuals, and product development. These three categories of innovation can also be applied simultaneously to create a "culture" where individuals are continually seeking to be innovative and create enhanced product outcomes. The meaning of innovation has evolved with U.S. Federal funding agencies as well. For example, consider the National Science Foundation (NSF), one of the premier research funding agencies in the United States that funds 24 percent of all federally supported basic research conducted by colleges and universities in the United States each year [7].

For many years NSF largely focused on funding only basic research rather than funding applied research and technology transition. Now the NSF's funding goals are extending beyond basic research to support various aspects of groundbreaking applied research and the transition of research outcomes into useful products, services, and technologies. There's a good reason for this change in focus. Historically, it was thought that it could take up to 50 years for the knowledge learned from basic research to be applied to products and services. However, as the pace of change itself continues to increase, the speed of technology and new development has compressed the time it takes to move basic research from reaction to knowledge to actual application. The NSF reflects this shift quite powerfully in its want to now fund more applied research.

The quick transition of the NSF's innovation core and its desire to swiftly convert new knowledge into new products and services is solid evidence of change.

## *Discussion of I-Corp program and related NSF initiatives*

America's affluence grew in part from the capability to profit economically on groundbreaking developments from science and engineering research. At the same time, a well-informed, imaginative labor force has maintained the country's international leadership in significant areas of technology. These essential discoveries and competent labor force resulted from substantial, incessant investment in science and engineering. A strong capability for leveraging essential scientific discoveries into influential engines of innovation is necessary to maintain our competitive edge in the future.

The NSF supports fundamental research and education in science and engineering. NSF's dual role, distinctive among government agencies, results in new knowledge and paraphernalia as well as a competent, groundbreaking workforce. These corresponding building blocks of innovation have led to innovatory high-tech advances and completely new industries.

Through this program, NSF seeks to hasten the improvement of new technologies, products, and processes that arise from elementary study. NSF investments will advantageously strengthen the innovation ecosystem [8] by addressing the challenge inbuilt in the early stages of the innovation process. This solicitation will support partnerships that are designed to triumph over scores of obstacles in the path of innovation.

### *Program description*

The objectives of this program are to encourage translation of fundamental research, to facilitate collaboration between the academic world and business, and to train students to comprehend innovation and entrepreneurship. The rationale of the NSF I-Corps program is to spot NSF-funded researchers who will obtain extra support—in the form of mentoring and funding—to hasten the conversion of knowledge derived from essential research into up-and-coming products and services that can attract successive third-party funding.

## *About the NSF*

The NSF is an autonomous federal agency created by the National Science Foundation Act of 1950, as amended (42 USC 1861-75). The act states the function of the NSF is "to promote the progress of science; [and] to advance the national health, prosperity, and welfare by supporting research and education in all fields of science and engineering" [7].

NSF funds research and learning in most fields of science and engineering through grants and cooperative agreements to more than 2000 colleges, universities, K-12 school systems, businesses, informal science organizations, and other research organizations all over the United States. The foundation accounts for about one-fourth of federal support to educational institutions for essential research. NSF receives in the region of 40,000 proposals each year for study, learning, and training projects, of which roughly 11,000 are funded. In addition, the foundation receives thousands of applications for graduate and postdoctoral fellowships. The agency operates no laboratories itself but does support national research centers, user facilities, certain oceanographic vessels, and Arctic and Antarctic research stations. The foundation furthermore supports joint research between universities and industry, U.S. participation in global scientific and engineering efforts, and educational activities at every academic level [7].

The role of creativity and innovation has changed our nation because now we are pushing more to see these new developments converted into new products and services, and the driving factor in accomplishing this is leadership. There is even more accountability in terms of wanting to understand what has been done with research funding for over the past several years.

Generally, Americans convey extremely favorable attitudes toward science and technology (S&T). In 2001, overpowering majorities of NSF survey respondents agreed with the following statements:

- "Science and technology are making our lives healthier, easier, and more comfortable" (86 percent agreed and 11 percent disagreed).
- "Most scientists want to work on things that will make life better for the average person" (89 percent agreed and 9 percent disagreed).

- "With the application of science and technology, work will become more interesting" (72 percent agreed and 23 percent disagreed).
- "Because of science and technology, there will be more opportunities for the next generation" (85 percent agreed and 14 percent disagreed) [9].

In addition, Americans give the impression to have more positive attitudes toward S&T than their counterparts in the United Kingdom and Japan [10].

Despite these positive indicators, a sizable segment, although not a majority, of the public has some reservations concerning science, especially technology. For example, in 2001, approximately 50 percent of NSF survey respondents agreed with the following statement: "We depend too much on science and not enough on faith" (46 percent disagreed). In addition, 38 percent agreed with the statement: "Science makes our way of life change too fast" (59 percent disagreed) [11].

## *Public attitudes toward federal funding of scientific research*

All indicators point to general support for government funding of essential research. In 2001, 81 percent of NSF survey respondents agreed with the following statement: "Even if it brings no immediate benefits, scientific research that advances the frontiers of knowledge is necessary and should be supported by the Federal Government" [12]. The level of agreement with this statement has consistently been in the 80 percent range. In 2000, 72 percent of U.K. residents agreed with the statement, as did 80 percent of Japanese residents (in 1995).

These differences in the measure of public support worldwide for basic research are notable. This may be attributed to the increased expectations in terms of transitioning science to technology and innovations. The result is people expect for basic research to more readily provide benefits to society and in fact, in 2001, 16 percent disagreed with the statement completely. This suggests that expect immediate benefits from basic research and this trend of expectation have continued.

Although there is strong evidence that the public supports the government's investment in basic research, few Americans are able to name the two agencies that provide most of the federal funds for this type of research. In a recent survey, only 5 percent identified the National Institutes of Health (NIH) as the agency that "funds most of the taxpayer-supported medical research performed in the United States," and only 3 percent named NSF as "the government agency that funds most of the basic research and educational programming in the sciences, mathematics and engineering" [13].

In addition, those with more positive attitudes toward S&T were more likely to express support for government funding of basic research. In 2001, 93 percent of those who scored 75 or higher on the Index of Scientific Promise agreed that the federal government should fund basic scientific research compared with only 68 percent of those with relatively low index scores [13].

In 2001, only 14 percent of NSF survey respondents thought the government was spending too much on scientific research; 36 percent thought the government was not spending enough, a percentage that has grown steadily since 1990, when 30 percent chose that answer [14]. Men are more than likely than women to say the government is spending too little in support of scientific research (40 percent versus 33 percent in 2001).

To put the response to this item in perspective, at least 65 percent of those surveyed thought the government was not spending enough on other programs, including programs to improve health care, help senior citizens, improve education, and reduce pollution. Only the issues *space exploration and national defense* received less support for increased spending than scientific research.

In 2001, 48 percent of those surveyed thought spending on space exploration was excessive, the highest percentage for any item in the survey—and nearly double the number of those who felt that the government was spending too much on national defense [15]. In contrast, the latter has been falling steadily, from 40 percent in 1990 to 25 percent in 2001.

## *Defining innovation*

Definitions of innovation differs but the general thread among these definitions is that innovations present a new or better product, service, or resource that adds "value" to those seeking it. The ETI study conducted 60 interviews that revealed common characteristics of innovators. These characteristics are shown in Figure 2.1.

A prevailing aspect of innovation is team interaction or team activities. For these teams to be effective, they are often managed by a technical person with detailed knowledge of the proposed innovation. In these situations, it is imperative that the team leader understands how to inspire, motivate, and lead the team as they move toward a useful innovation.

When innovators were asked to describe characteristics of innovations or innovative products, the following characteristics emerged:

- **Innovation Provides Societal Value**
  The interviewees felt powerfully that innovations must offer societal value. The innovation must be supportive to society. It's great if

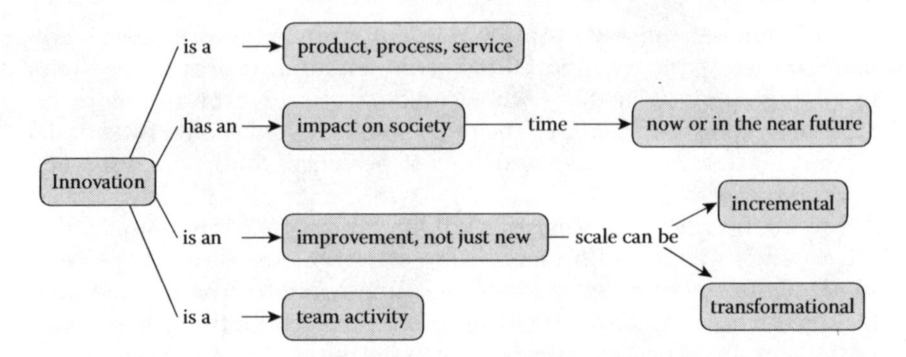


***Figure 2.1*** Characteristics of innovation. (From National Academy of Engineering, *Educate to Innovate*, National Academies Press, Washington, DC, 2015, p. 11. With permission.)

one makes an invention, but it's even better if the invention can be used to develop individual lives. Part of the importance of an innovation is connected to timely adoption. It should be helpful in the near future. In truth, unless an innovation is in fact used by society, it cannot be called an innovation: R. Graham Cooks* [16] warned innovators into believing that all they do is collectively meaningful or useful: In other words, "if you feel that you have some fondness for innovation, then the big danger is that you'll convince yourself even in cases where the work is trivial or doesn't have the implications that you hoped it would have." Robert Dennard† [16] agreed: "Lots of inventions aren't innovations. I have 62 patents and only one or two are actually being used, and if it's not used, it's really not innovating very much. So innovation's a breakthrough, something that's really useful and it doesn't have to be patentable, even."

- **Innovation is an Improvement**
  Innovations are naturally seen as "something new." Nevertheless, all the interviewees and workshop participants accentuated that innovations are improvements, not necessarily just new.

* Cooks is a distinguished professor of chemistry at Purdue University. His interests involve construction of mass spectrometers as well as studies of their fundamentals and applications. His work on ionization methods has contributed to the ambient method of desorption electrospray ionization, which is used in tissue monitoring, forensics, and pharmaceutical applications.

† Dennard is a retired IBM Fellow known for inventing dynamic random access memory (DRAM) and formulating the scaling theory, making it possible to miniaturize the channel lengths of metal oxide semiconductor field-effect transistors, or MOSFETs, down to just nanometers.

Laurie Dean Baird* [16] gives details of her approach to tell the difference of the value of an innovation: "If I look at something that's new and ask 'Is this innovative?' then I ask 'How was this problem solved before? What was the industry standard and how is this different?' And if the answer is that, in addition to being new (the problem or solution), it takes the hassle out of something (i.e., it improves life), then it is innovative."

"I don't see innovation being the introduction of something [that is just] new." Tim Cook† says, "There are many things new every day, and I wouldn't say they all are innovative. I think to be innovative, something has to be better than the predecessor product, materially better, not just a small percentage better" [16].

In terms of the level of improvement, innovations can be transformational, for instance, creating large-scale changes in the way technology is used or thought about. Mary Lou Jepsen‡ [16] said, "I think of innovation as doing some transformative work in an area or in a combination of areas that trail blazes in a way that people recognize has moved the ball forward … in a way that is a leap."

But it is not compulsory that every innovation be pioneering or radically change the world. Bernard Meyerson§ [16] referred to "continuous innovators": "The danger is there are other types of innovators that are just as necessary, what I call the continuous innovators. These are the guys who come to work every day and make it 5 to 10 percent better, and there's a terrible undervaluation of that."

- **Innovation Occurs at the Interfaces of Different Disciplines**
  Innovators in all the areas represented, that is, academia, large companies, small businesses, and the arts acknowledged that innovation occurs at the edge of disciplines and necessitated the synthesis

* Baird is a strategic consultant in media and entertainment and focuses on emerging technology, social practices, and business models in the changing media landscape. She is a research fellow at the Futures of Entertainment and a strategic consultant at the Georgia Tech Institute for People and Technology.

† Tim Cook is Apple's CEO and serves on its board of directors. As CEO, he has encouraged greater collaboration and creativity among Apple's team, which is widely regarded as the most innovative in the world. Before being named CEO in August 2011, he was Apple's COO and was responsible for the company's worldwide sales and operations. In his time at Apple, he has helped improve conditions for workers who make the company's products, and is today leading a companywide effort to use 100 percent renewable energy at all Apple facilities.

‡ Jepsen is head of the Display Division at Google X. She is also the founder and former CEO of Pixel Qi, a manufacturer of high-performance, low-power, sunlight-readable screens for mobile devices, and cofounder and former CTO of One Laptop per Child.

§ Meyerson is an IBM Fellow, IBM's vice president for innovation, and drives corporate initiatives in IBM's Corporate Strategy Function. He has been part of the IBM family since 1980, led the development of silicon germanium and other high-performance semiconductor technologies, and held a wide range of positions in broad executive management.

of knowledge from dissimilar fields. Yo-Yo Ma* [16] captured this aspect using the concept of the edge effect from ecology: "If you think about where new ideas can come from, you need proximity to density, and if you're at the edge of something you see both sides; you already see over the wall. You could be part of one ecosystem, but you actually are constantly interacting with another ecosystem, and so you see the possibility of what another ecosystem can bring. And ... if the center uses the knowledge at the edge, the center does benefit."

- **Teamwork is Important to the Process of Innovation**
  Innovation is the effect of joint effort, a point frequently made by the innovators. And it relies on the work of the team as a whole, not the work of one key innovator and other "supporters." Ivan Seidenberg† [16] observed: "I get comfort in knowing that life is cumulative, innovation is cumulative, and it's not individual. Let's take some of the greatest examples: Let's start with the example everybody's using right now, and I knew him well. Steve Jobs was a genius, but he didn't invent the computer; he didn't invent anything that went into the iPhone, but he made it all work together ... so what did he invent? Take another example: Bill Gates had enough common sense and enough vision to know that PCs couldn't talk to each other, so he built operating systems to make them talk to each other, but along the way, they didn't work very well when they first came out with them. They (Jobs and Gates) needed a full team and with their superior insights and innovative spirit, they made something bigger than any one person could have made. So all I'm getting at is that there's really no one innovator who can innovate all alone. I can't think of any one person that gets it all right. Is there anybody? Is there anybody in the literature that gets it right the whole time?"

- **Innovation is Part of an Invention–Value Continuum**
  Innovation is part of a field between invention and worth. Innovators may start with a discovery and then innovate to

* Ma is an American cellist. He began to study the instrument at the age of 4, and his discography numbers over 90 albums, including more than 17 Grammy Award winners. One of his goals is the exploration of music as a means of communication and as a vehicle for the migration of ideas across cultures throughout the world. To that end, he founded Silkroad, a nonprofit organization that, through performance and the creation of new music, cultural partnerships, education programs, and cross-disciplinary collaborations, seeks to create meaningful change at the intersection of the arts, education, and business.

† Seidenberg is the former chair and CEO of Verizon Communications. He worked in the communications industry for more than 45 years, and is known for steering the merger of Bell Atlantic and NYNEX in 1997 and the Bell Atlantic merger with GTE in 2000. He also led efforts to form Verizon Wireless.

generate value from it, or start with a problem and solve it innovatively. Innovation was portrayed as the use of inventions to real-world needs. Innovation can also be driven by the impression of marketability or attempt to solve a problem. As Robert Fischell* [16] said, "Sometimes we see an invention and then we can apply it to another thing, but that doesn't happen very often. Most times, we hear about something and it occurs to us that the way they're doing it is not good, and so we innovate a better way."

Analysis of the 60 innovators' observations disclosed that innovation is an enhanced product, process, or service that profits society in a timely and, sometimes, transformational manner. It is a team activity at the meeting point of diverse fields, bringing as one diverse ideas, skills, and/or methods to result in the production of value.

## *Types of innovation*

Innovation applications are commonly applied to either a product, a process, or a service. To additionally comprehend how this is done, let's reflect on three categories of innovation:

1. **Product Innovation**
   Product innovation is about making valuable changes to material products. Interrelated terms that are frequently used interchangeably comprise *product design, research and development,* and *new product development* (NPD). All of these terms proffer a particular viewpoint on the degree of alteration to products.

   Well-known organizations characteristically have a collection of products that must be incrementally enhanced or adjusted as problems are recognized in-service or as new requirements emerge. It is imperative that they also work on add-ons to the product families. One of the major actions of the product design team is the work it carries out on next-generation products or new models of products. They might also work on designing far-reaching new products or new core products that enlarges the portfolio considerably and frequently involve drastically new processes to produce them. These new core products idyllically present the organization the possibility of major increases in revenue and growth, which can also create the potential of short-term monopoly in the market.

* Fischell is a physicist, inventor, and holder of more than 200 U.S. and foreign medical patents. He has had two pioneering careers. His current career is characterized by forming several biotechnology companies to develop and refine his inventions and innovations.

The product development process for next-generation and new core products according to Cooper follows a familiar cycle in most organizations:

a. Ideation
b. Preliminary investigation
c. Detailed investigation
d. Development
e. Testing and validation
f. Market launch and full production [17]

All of these steps involve communication with customers, who might take part in idea creation and element recognition. Key performance criteria in the design process revolve around the following:

a. Time to market
b. Product cost
c. Customer benefit delivery
d. Development costs [18]

These standards can be traded off against one another. For instance, development costs can be traded against time to market, customer benefits can be traded against product costs, and so on. Three blueprint systems have ascertained themselves as providing a management system for efficient product innovation: phase review, stage gate, and product and cycle-time excellence (PACE).

a. **Phase Review**

   This technique splits the product development life cycle into a sequence of different phases. Every phase encompasses a body of work that, once finished and evaluated, is dispensed over to the next phase. No consideration is paid to what may or may not occur in the succeeding phases, principally for the lack of knowledge or exclusive focus on the job in the existing phase. The phase review technique is a chronological rather than a simultaneous product design method, that is, each phase is accomplished and concluded before the commencement of the next phase.

b. **Stage Gate**

   This technique is a simultaneous product design procedure that follows a prearranged life cycle from idea creation to market commencement [17]. The stages in this technique are first and foremost cross-functional. Stage gates appear at the end of each stage, where a design evaluation takes place. Each stage gate evaluates the decided deliverables for completion at the conclusion of the stage, a checklist of the standard agreed for each stage, and a choice about how to advance from a particular stage.

c. **PACE**

   This method is concerned mainly with enhancing product improvement strategies [18]. The technique connects product strategy with the general strategy and goal of the organization. A key element is positioning of the voice of the customer all through the product design procedure. Strategies are divided into six product strategic thrusts: expansion, innovation, strategic balance, platform strategy, product line strategy, and competitive strategy. Product innovation methods and processes are one element in an organization's mission to create value for customers.

2. **Process Innovation**

   Process innovation can be observed as the launching of a new or considerably enhanced method for the construction or delivery of production that append value to the organization. The term *process* refers to an interconnected set of actions designed to convert inputs into a specific result for the customer. It implies a strong prominence on how work is done within an organization rather than what an organization does [19].

   Processes recount every operational action by which value is presented to the end client, such as the purchase of raw materials, production, logistics, and after-sales service. The process innovation in the 1970s and 1980s gave Japanese manufacturing a viable advantage that permitted them to take over some international markets with cars and electronic goods. Likewise, process innovation has permitted organizations such as Dell and Zara to achieve competitive advantage by offering higher-quality products, delivered faster and more proficiently to the market than by the competitors. By focusing on the resources by which they transform inputs, such as raw materials, into results, such as products, organizations have achieved efficiencies and have added importance to their production. Process innovation permits some organizations to contend by having a further proficient value chain than their rivals have.

   Process innovation has resulted in organizational enhancement such as lower stock levels; quicker, additional flexible production processes; and more responsive logistics. Organizations can develop the competence and value of their processes with a huge array of diverse enablers. Even though the use of these enablers is dependent on the organizational framework, many present the possibilities for improved process performance. The application of technology such as robotics, enterprise resource planning systems, and sensor technologies can change the process by decreasing the price or variation of its output, improving safety, or decreasing the throughput time of the process.

3. **Service Innovation**

   Service innovation is concerned with making changes to intangible products. Services are frequently linked with work, play, and recreation. Examples of these types of service consist of education, banking, government, recreation, entertainment, hospitals, and retail stores. In the past decade, an enormous amount of knowledge-based services has been accessible through websites. These services involve intangible products, have a high quantity of customer dealings, and are typically set in motion on demand by the customer. Defining a service can be to some extent problematic. Some define service as a sequence of overlapping value-creating activities.

Others define service in terms of performance, where customer and provider coproduce value. There are three categories of service operations:

a. Quasi-manufacturing (e.g., warehouses, testing labs, recycling)
b. Mixed services (e.g., banks, insurance, realtors)
c. Pure services (e.g., hospitals, schools, retail)

Services can without a doubt involve products that form a comprehensive part of the product life cycle, from preliminary sales to end-of-life recycling and clearance. Service business in areas such as finance, food, education, transportation, health, and government make up most organizations in any economy.

These organizations as well require innovation incessantly so that they can enhance levels of service to their customers. A key characteristic of a service is a very high level of communication with the end user or customer. The customer is often not capable of separating the service from the person delivering the service and so will make quality postulation based on impressions of the service, the group delivering the service, and any product delivered as part of the service. An additional feature of some service organizations is that their product may be perishable; consequently, the product must be consumed as soon as possible following purchase. Consequently, the timing of the delivery and customer opinion of quality are vital to success.

The notion of service quality is of particular significance. Service quality is a function of numerous factors including the uniqueness of offerings, intangibilities such as customized customer contact or perishable manufacture and a continued capacity for innovations of the service. Another important driver of service innovation comes from the possibilities afforded by the new information technology podium, predominantly the Internet. The Internet is a priceless resource on which new service associations between organizations and their customers are being developed every day.

## *Innovation and entrepreneurship*

If innovation is successful, the expected outcome is the transitioning of these new products, processes, or services into useful products that people are willing to pay for in the United States and globally.

Although innovation and entrepreneurship are related, many caution the intent of focusing too much on entrepreneurship in the initial stages of the creative aspect of innovation. This perspective believes that entrepreneurship should be a natural outcome of entrepreneurship but should not be the initial focus.

> It's really important to lead with "innovation" and have it evolve into "entrepreneurship" because innovation is the large end of the funnel that appeals to and actually requires participation by a much broader audience. Non-business, non-engineering, and non-STEM people are every bit as important to include in that innovation process because the process is not as rich and has inferior outcomes without that diversity. [20]

In order to see this type of innovation systematically realized, engineering leaders must understand principles that should be integrated into the creative process to produce effective innovations.

The terms *entrepreneurship* and *innovation* are over and over again used interchangeably, nevertheless this is deceptive. Innovation is frequently the starting point on which an entrepreneurial business is built for the reason of the competitive advantage it offers. On the contrary, the act of entrepreneurship is simply one means of bringing an innovation to the marketplace. Technology entrepreneurs regularly decide to build a startup company for a technological innovation. This will offer financial and skill-based resources that will take advantage of the chance to grow and commercialize the innovation. Once the entrepreneur has set up a business, the focal point shifts in the direction of its sustainability, and the best way to attain this is through managerial innovation. Nonetheless, innovation can be conveyed to the market by ways other than entrepreneurial startups; it can also be subjugated through well-known organizations and deliberate alliances between organizations.

---

**Case study: Charles Dow**

In 1896, Charles Dow created the Dow Jones Industrial Average in order to provide a snapshot of the U.S. economy through the stock market. There were 12 companies on Dow's original list: American Cotton Oil, American Sugar, American Tobacco, Chicago Gas, Distilling & Cattle Feeding, General Electric (GE), Laclede Gas, National Lead, North American, Tennessee Coal and Iron, U.S. Leather pfd., and U.S. Rubber. Of all of those companies, which were financial leaders at the turn of the 20th century, there is only one you might recognize that is still in business today: General Electric.

What is the key to GE's century-long tenure? Product innovation. According to business researchers, Heath Downie and Adela J. McMurray, "The consistency of GE's commitment to product innovation was made possible by the steadiness of the company's leadership." Even during the Great Depression, GE found a way to allocate diminishing financial resources to its research and development initiatives.

Today, GE has taken their commitment to innovation even further, crowdsourcing both internally and externally to drive advancements in several industries. In fact, GE has an Open Innovation Manifesto, in which they state:

> We believe openness leads to inventiveness and usefulness. We also believe it's impossible for any organization to have all the best ideas, and we strive to collaborate with experts and entrepreneurs everywhere who share our passion to solve some of the world's most pressing issues [...] We'll never stop experimenting, collaborating and learning—we'll get smarter as we go, and the Global Brain will evolve and grow with us.

GE has a hand in advancing just about every engineering industry you can think of, such as aviation, software, consumer goods, water and wastewater, power and energy, transportation, and healthcare, to name a few. Named "America's Most Admired Company" in a poll conducted by *Fortune* magazine and one of "The World's Most Respected Companies" in polls by *Barron's* and the *Financial Times*, the quality work GE has done for the planet has not gone unnoticed, and their leadership is extremely dedicated to quality and innovation.

Take, for instance, Deb Frodl, global executive director of GE Ecomagination. Ecomagination is a business initiative designed by GE to develop innovative solutions to environmental challenges while driving economic growth. In an interview with Cleantech Group, Frodl said:

> Innovation is the foundation for Ecomagination and we have really developed a lot of solutions that solve complex problems for a multitude of industries. [...] Ecomagination has really been the catalyst within GE to step outside and get those ideas and that outside innovation moving forward.

In 2016, GE announced it would be relocating its corporate headquarters to Boston, Massachusetts, in part to enable GE to place additional emphasis on digital industrial innovation. This is further proof of the company's commitment to innovation and its leaders' push to improve access to a more innovative workforce and relocate to a better environment for innovation.

**QUESTIONS**

1. What engineering leadership characteristics were demonstrated in this case study?
2. How were these characteristics applied to deliver the intended impact?
3. What was the technical, societal, or environmental impact due to the application of the leadership principle(s) identified above?

---

**Real profiles in engineering leadership**

**Name:** Bryan Lin

**Current position or field of expertise:** Quality Engineer, Disney

**What I like most about my position:** The opportunity to freely devise, manage, and improve the quality systems for Disney's MagicBand development, production, and fulfillment processes.

**Who or what has made a difference in my career:** My parents always challenged me to think logically, excel in what I did, and to find creative solutions to problems whenever they arose. Combined with discovering the game of chess at an early age, I developed a strategic planning mindset that led me to my field in industrial engineering and has served me ever since.

As someone who has always been more introverted by default, my founding and subsequent leadership of the All-male a cappella group Crescendudes at the University of Florida from 2005 to 2009 shaped the foundation of my leadership style and made a distinct impact in my personal development. Nothing highlights the criticality of developing relationships better than leading a group of guys to volunteer two hours of their time three times a week to rehearse with other guys. Titles mean very little—it's what you do and how you do it that counts. I'm glad I put myself out there; they were some of the most fun years of my life, and I learned a lot about democratic leadership, buy-in, and relationships from my experiences with those guys. I still work with them from time to time even today.

Lastly, I must also thank my mentors and the people who were willing to take the chance, particularly at the beginning, who gave me the opportunity to learn, grow, and prove myself in my various roles that have led me to where I am today.

**I felt like quitting when:** I am at my best under pressure, and more often than not am fairly invested by the time I have landed myself in that kind of situation. But like anyone else, continuous stress, feeling unappreciated, and setbacks can be discouraging and can burn you out. Take a break from time to time. It's important to maintain a good work/life balance. More often than not it's not the company but good or bad management that makes a huge difference in the quality of a work environment. Seek out new opportunities if you need to. Similarly, if you are that management, make sure that you are creating a positive environment for your employees to work.

**My strategies for success are:**

- *Be authentic*: It's so much easier to be honest and direct than to play mind games. Everyone will also appreciate that your integrity can be counted upon.
- *Help others because you can and want to, not because you must*: They will remember it, and all the more so if you have no particular obligation. Word spreads, and who knows, maybe it will come back to help you someday. Even if it doesn't, you made someone's life a little better.
- *Time management*: if it takes less than five minutes, do it NOW. For everything else, identify what can be taken care of now and what is long term. Take care of the ones you can deal with today and save the rest for worrying about another day. To-do lists are a lot less stressful when you can clear off all the small tasks that make the list seem much tougher than it is.
- *Planning ahead*: Nothing ever goes exactly as planned. The more contingencies you have accounted for, the more likely things are to go smoothly. Have a plan B, a plan C. But also leave room for flexibility and adaptability in the plan.
- *Be realistic*: Hope for the best, prepare for the worst, and expect to get something in between.

**I am excited to be working on:** The next generation of the Disney magic experience and helping make memories that will last a lifetime!

**My life is:** I'm very lucky and grateful to have been presented the opportunities to get to where I am today. Many of my generation have not been so fortunate and not for lack of intelligence or effort on their part. Many people didn't have the guidance, the financial cushion, the spare time, or the opportunities that I had. They have fallen through the cracks and may never have the kind of future that they and their parents had hoped for, through no particular fault of their own. But perhaps with a little more empathy from each of us and a stronger sense of collective social responsibility, we can extend to them and those yet to come, the opportunity to change their lot in life and contribute to making a brighter world for all of us. Who knows when it will be our turn to ask for help. Today you, tomorrow me.

---

## References


1. National Academy of Engineering. *Educate to Innovate.* Washington, DC: National Academies Press, 2015.
2. U.S. Department of Education, National Center for Education Statistics. *Highlights from PISA 2006: Performance of U.S. 15-Year-Old Students in Science and Mathematics Literacy in an International Context.* 2007. Available at: http://nces.ed.gov/pubs2008/2008016.pdf, accessed on October 12, 2016.
3. National Commission on Excellence in Education. *A Nation at Risk.* 1983. Available at: http://www.ed.gov/pubs/NatAtRiskrisk.html, accessed December 1, 2010.
4. The National Academies Press. *Rising above the Gathering Storm, Revisited: Rapidly Approaching Category 5.* 2010. Available at: https://www.nap.edu/catalog/11463/rising-above-the-gathering-storm-energizing-and-employing-america-for.
5. The White House. *Remarks by the President at the National Academy of Sciences Annual Meeting.* The White House. 2009. Available at: https://www.whitehouse.gov/the-press-office/remarks-president-national-academy-sciences-annual-meeting, accessed on October 12, 2016.
6. The White House. *President Obama Launches 'Educate to Innovate' Campaign for Excellence in Science, Technology, Engineering & Math (Stem) Education.* 2009. Available at: http://www.whitehouse.gov/the-press-office/president-obama-launches-educate-innovate-campaign-excellence-science-technology-en, accessed December 1, 2010.
7. National Science Foundation. *Fact Sheet January 7, 2016.* P.1. Available at: https://www.nsf.gov/about/congress/reports/nsf_factsheet.pdf, accessed on October 13, 2016.

8. NSF Directorate for Engineering version. *The Role of the National Science Foundation in the Innovation Ecosystem.* Available at: http://www.nsf.gov/eng/iip/innovation.pdf, accessed October 13, 2016.
9. Gallup Organization. *Science & Engineering Indicators—200 "Only One in Four Americans Are Anxious About the Environment,"* Poll Release, Princeton, NJ, 2001. Available at: https://wayback.archive-it.org/5902/20150819162635/http://www.nsf.gov/statistics/seind02/c7/tt07-04.xls, accessed on October 13, 2016.
10. Prime Minister's Office. 1995 "Public Opinion Survey of Relations of Science and Technology to Society." Survey taken in February 1995, and "Public Opinion Survey of Future Science and Technology." 2001. Survey taken in October 1998. *S&T Today* 14(6): 12. Tokyo: Japan Foundation of Public Communication on Science and Technology.
11. Gaskell, G., and Bauer, M.W. (editors). *Biotechnology* 1996–2000, National Museum of Science and Industry (U.K.) and Michigan State University Press. *Science & Engineering Indicators*—200. Available at: https://wayback.archive-it.org/5902/20150818093236/http://www.nsf.gov/statistics/seind02/c7/fig07-11.xls, accessed on October 13, 2016.
12. Research! America. *Research! America. Poll Data Booklet.* Vol. 2. Alexandria, VA: Research! America, 2001.
13. National Science Foundation. *Survey of Public Attitudes toward and Understanding of Science and Technology.* Arlington, VA: National Science Foundation, 2001.
14. Pew Research Center for the People and the Press. *America's Place in the World II: More Comfort with Post-Cold War Era.* 1997. Available at: http://www.people-press.org/1997/10/10/americas-place-in-the-world-ii/, accessed on October 13, 2016.
15. Carlson, D.K. Public Views NASA Positively, but Generally Disinterested in Increasing Its Budget. *Gallup News Service.* Poll Analyses. Available at: http://www.gallup.com/poll/releases/, accessed on March 8, 2001.
16. Bement, A., Jr., Dutta, D., and Patil, L. 2015. *Educate to innovate: Factors that influence innovation: Based on input from innovators and stakeholders.* National Academies Press.
17. Cooper, R.G., and Kleinschmidt, E.J. New product performance: What distinguishes the star products? *Australian Journal of Management* 2000; 25(1): 17–46.
18. McGrath, M. *Setting the Pace in Product Development. A Guide to Product and Cycle-Time Excellence.* Revised Edition. Stoneham, MA: Butterworth-Heinemann, 1996.
19. Davenport, T.H. *Process Innovation: Reengineering Work through Information Technology.*
20. National Academy of Engineering. "What Is Innovation?" In: *Educate to Innovate: Factors That Influence Innovation: Based on Input from Innovators and Stakeholders,* Academy of Management 7(2), May 1993. Washington, DC: The National Academies Press, 2015. doi: 10.17226/21698.

# chapter three

# Leadership within an organization

To be effective as a leader, it is important to understand how to add value in support of the vision and mission of an organization. Thus, an individual must gain some knowledge of organizational leadership within the environment, the specific organizational culture, and finally the organizational structure.

## Understanding organizational leadership

Part of being able to correctly lead within an organization depends on understanding the functioning of the particular organization. Organizations are far more dynamic than they once were; people no longer stay at the same company for their entire careers. It is essential to quickly adapt to and understand the needs of your organization in order for both you and it to grow. Thus, it is even more important that a clear perspective be gained early in one's tenure at an organization. Generally, individual leaders are established throughout organizations to support the overall organizational objectives and goals. According to Zaccaro and Klimoski [1], the primary areas of organizational leadership that have some degree of consensus in the literature can be narrowed down to four principles. These broad principles include the following:

1. **Organizational Leadership Involves Processes and Proximal Outcomes (such as Worker Commitment) that Contribute to the Development and Achievement of Organizational Purpose**
   Leadership positions are setup in the work place to enable organizational subunits to achieve the reasons for their existence within the larger system. The purpose of an Organizational Leadership is to set a direction for collective actions. The process of leadership is aimed at defining, identifying, establishing, or translating organizational directions for their followers and aiding or enabling the organizational processes that should result in the accomplishment of this function. Organizational purpose and direction is described in several ways, including through mission, vision, strategy, goals, plans, and tasks. The operation of leadership is inextricably tied to the constant growth and accomplishment of these organizational goals.

This viewpoint of leadership is a purposeful one, denoting that leadership is at the service of combined effectiveness [2]. Recounting a similar approach to team leadership, Hackman and Walton [3] bickered that the leader's "'main job is to do, or get done, whatsoever is not being satisfactorily handled for group needs.' Thus by whatever means necessary, if a leader manages to make certain that all functions significant to both task completion and group preservation are sufficiently taken care of, then, the leader has done his or her job well." These adaptations can be made whether individuals are leading single group, multiple groups merged into a department or a division, the organization as a whole, or a multinational organization. This crucial component of organizational leadership in addition means that the achievement of the collective as a whole is a primary factor for leader effectiveness.

Functional leadership is normally not defined by any particular set of behaviors but instead by a common leadership strategy that is predetermined but with the flexibility to adapt to diverse circumstances. That is, the importance switches from "what leaders should do" to "what needs to be done for effective performance" [3]. Consequently, leadership is described in terms of those activities that advance team and organizational goal accomplishment by being receptive to relative demands [4].

The organizational purpose and direction are usually framed by the top leaders of an organization. The essentials of leadership performance that originate from the organizational context become tangled in this commitment as well as in the content of organizational tracks and the intricacy of the top leader's operating background. This necessitates significant cognitive capital to put together the frame of reference that provides the basis for organizational policy. In the same way, organizational goals and strategies need to be receptive to the needs of manifold stakeholders and constituency, communities and constituencies, associated with the organization. Lastly, a little-noted observation concerning organizational goal setting is that when such guidelines are formed, they reproduce in part the senior leader's personal and self-defined (occupation) imperatives. As a consequence, when leaders build up and execute organizational strategies, they do so from and within the framework of these and the other imperatives discussed here.

2. **Organizational Leadership Is Identified by the Application of Nonroutine Influence on Organizational Life**

   Leadership does not subsist in the everyday actions of organizational work. Instead, it arises as a reaction to, or in expectation of, nonregular organizational actions. This important factor was suggested by

Katz and Kahn [5], who measured "the essence of organizational leadership to be the influential increment over and above mechanical compliance with the routine directives of the organization."

Noncustomary events can be described as any circumstances that comprise a possible or definite impediment to organizational objective development. Therefore, organizational leadership can be seen as large- (and small-scale) societal problem-solving, where leaders are raising the nature of organizational problems; developing and accessing a possible way out; and scheduling, executing, and observing chosen solutions inside difficult social domains [1,2]. This is not to propose that leadership as a social problem-solving is unavoidably imprudent. The border management approach which is largely consigned to organizational leaders according to Katz and Kahn [5] requires that leaders be accustomed to environmental events, interpreting and defining them for their followers while pro-actively providing mitigation techniques to address anticipated obstacles. Therefore, successful organizational leadership is quite proactive in its problem-solving.

This essential factor of leadership as involving nonroutine influence reflects two additional points. First and foremost, vital organizational leadership is further prone to be reflected in reaction to imprecise problems. In such circumstances, leaders could do with constructing the nature of the problem, as well as the stricture of likely solution strategies, before they can begin to work out resolutions to the problem. In clear problems, solutions are grounded typically in the knowledge of the leaders of previous similar situations. Such solutions are as well not likely to need significant large-scale alteration in organizational routine.

The next point is that leadership characteristically entails discretion and options in what solutions are suitable in specific problem domains [5]. Therefore, as Jacobs and Jaques [6] contend, "Leadership must be viewed as a process which occurs only in situations in which there is decision discretion. To the extent discretion exists, there is an opportunity for leadership to be exercised. If there is no discretion, there is no such opportunity."

Organizational activities that are totally specified by modus operandi or practice or are completely elicited by the situation do not generally require the interference of leaders. Such actions are probable to be determined as part of the organizational tenet or normative structures (even though, leadership is involved in the development of these structures). In its place, leadership is required by organizationally significant proceedings that present different explanations and by problems in which several solution paths are doable or mandatory solutions need to be executed. Persons in leadership positions are subsequently responsible for making the option that describes the ensuing collective responses.

The performance imperatives we highlighted in this sense can be taken to mean as representing a cluster of ill-defined optional problems or obligations requiring collective action for organizational success. For instance, the nature and rate of technical change can cause a number of problems to organizational leaders: how to gather and distribute information, how to comprehend the resulting flood of data, and how to achieve competitive benefits from high-tech advances in both production and human resource systems are just a few of these. In the same way, financial imperatives challenge executives to make and incorporate a range of long- and short-term calculated choices. Consequently, a functional or social problem-solving perspective of leadership is necessarily grounded in a related framework that presents essential performance imperative challenging organizational options.

3. **Leader Influence Is Grounded in Cognitive, Social, and Political Processes**
Most leadership definitions emphasize societal or interpersonal persuasion processes as major fundamentals. As a result, persuasion, the management of societal and political processes, and the utilization of social power are ever-present constructs in the leadership literature. In addition, as recommended by the problem-solving perspective, the implementation of efficient cognitive procedure is just as significant to leader efficacy. Usually, models of leadership, predominantly those in the psychosomatic literature, have focused on social procedures directed toward the implementation of solutions to organizational problems. A complete description of leadership must also include the cognitive procedures leaders use to plan joint action. A number of researchers in the organizational and management literatures have in reality emphasized the role of leaders in organizational sense making, where joint actions are given meaning through the leader's construal and cognitive modeling of environmental proceedings [6–8].

Beside these positions, Jacobs and Jaques [6] noted: Executive leaders "add value" to their organizations in large part by giving a sense of understanding and purpose to the overall activities of the organization. In excellent organizations, there almost always is a feeling that the "boss" knows what he or she is doing, that he or she has shared this information downward, that it makes sense, and that it is going to work. Such thoughts, delineated as frames of reference by Jacobs and Jaques [6,9], become a vital arbitrator of leadership influence in organizations. An organizational frame of reference is a cognitive depiction of the fundamentals and actions that encompass the leader's working environment. Such models enclose the outline of relations among these events and elements.

Therefore, the reason and justification for an expressed organizational policy become stuck in the causal associations construed by top executives as obtainable between the vital proceedings in organizational space. These frames of reference are not the only territory and responsibility of senior leaders. Every leader has to interpret their working environment and communicate that understanding to their constituency. Nevertheless, banking on the theory of necessary variety argued that the intricacy of the organizational causal map must match up to the intricacy of the working environment being patterned [9,10]. For this reason, the frames of reference or causal maps that senior leaders develop must be more intricate than those of leaders at lower organizational levels. In other words, executive leaders must consider many more causal elements, and the links among these elements at all levels of the organization.

It's not my intention to argue that the use of cognitive and societal leadership procedures is wholly self-determining. In several examples of efficient leadership, these procedures become inextricably intertwined. For illustration, functionally diverse teams (where members have various specializations in the organization) can assist leaders understand environmental uncertainty and decrease doubt. This is predominantly true in top management teams, where environmental intricacy is characteristically stronger than for lower-level leaders [11].

4. **Organizational Leadership Is Inherently Bounded by System Characteristics and Dynamics; That Is, Leadership Is Contextually Defined and Caused**

One predominantly powerful influence is the organizational level at which leadership happens. Besides the fundamental demands and work requirements of leaders changing at different levels [5,9], the hierarchical context of leadership has profound effects on the personal, interpersonal, and organizational choices that can be made, as well as the importance that a given choice might have. Evidently, a CEO stating a preference for a site for a new factory is different from the case of a department manager stating his or her preferences. Organizational-level matters overpoweringly, yet astonishingly, has been disregarded in all but a few leadership models in the literature.

What has been disputed regarding leadership at diverse organizational levels? Katz and Kahn [5] specified three distinct patterns of organizational leadership. The foremost pattern concerns the managerial use of accessible organizational structures to preserve efficient organizational operations. If challenges come up to disrupt these operations, existing organizational mechanisms and measures are used to resolve them. Indeed, Katz and Kahn note that "such acts are often seen as so institutionalized as to require little if any leadership." This leadership pattern takes place at lower organizational levels. The next leadership pattern, happening at middle organizational levels, comprises the embossing and maneuvering of official structural elements. Such events necessitate a two-way orientation by the leader (that is, toward both superiors and subordinates), as well as important human relations skills. The third pattern of organizational leadership, which takes place at the top of organizations, considers structural beginning or change in the organization as an indication of new policy formulations. Taken together, the sharing of separate leadership patterns across organizational levels that Katz and Kahn projected advocate important qualitative differences between the nature of junior and senior leadership. Related models specify differences across levels of organizational leadership that have been proposed in separate academic formulations by Jacobs and Jaques [9] and Mumford and Zaccaro [12].

## *Organizational culture*

Understanding an organization's culture is essential for success as a leader. Organizational culture is a pervasive and accepted "personality" of a corporation, academic institution, and even a community. For the purposes of professional environment, organizational culture can be defined as:

> A pattern of shared basic assumptions that a group has learned as it solved its problems of external adaptation and internal integration that has worked well enough to be considered valid and therefore, to be taught to new members as the correct way to perceive, think, and feel in relation to those problems. [13]

According to Edgar Schein [13], organizations do not accept a culture in a single day, instead it is created in due course of time as the workers go through a variety of changes, become accustomed to the outer surroundings, and resolve troubles, if any. They benefit from their past experiences and start to practice it every day consequently forming the culture of the workplace. The new workers also make every effort to adjust to the new culture and experience a stress-free life.

Schein presented culture as a sequence of assumptions a person makes about the group to which he or she belongs to (Figure 3.1). These assumptions

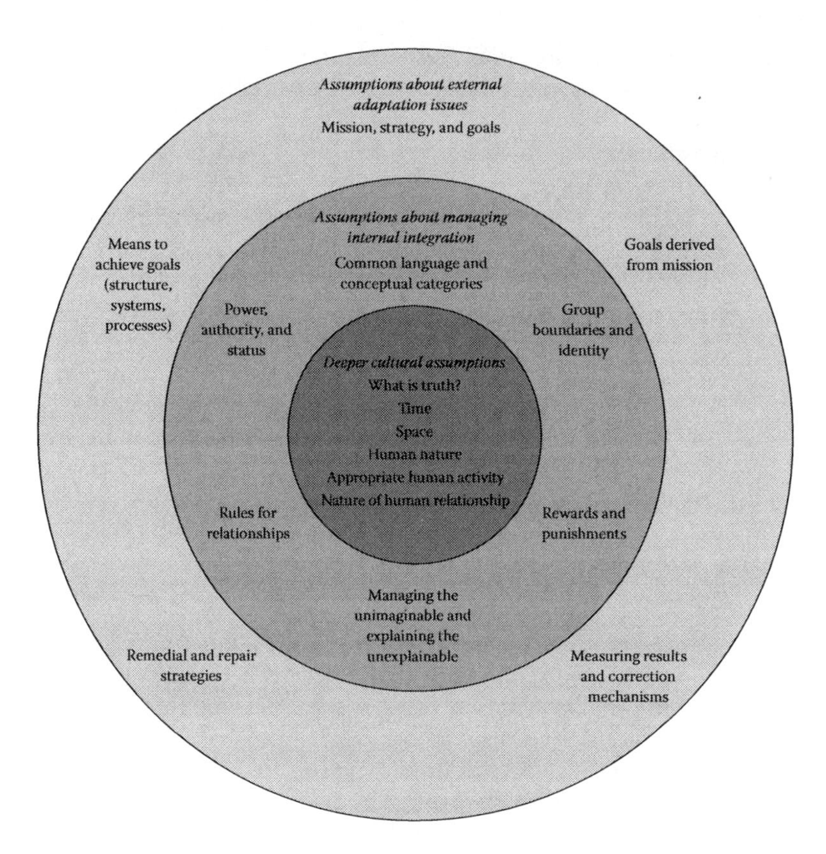


***Figure 3.1*** Culture according to Edgar Schein. (Adapted from Schein, E., *Organizational Culture and Leadership*, Wiley, 2010, pp. 589–592.)

are assembled into three levels, each level becoming additional complex to communicative and change. These assumptions can be seen through:

1. **Artifacts**
   The foremost level is the uniqueness of the organization which can be effortlessly viewed, heard, and felt by individuals jointly known as artifacts. The dress code of the workers, office furniture, facilities, conduct of the employees, mission and vision of the organization all come under artifacts and go a long way in deciding the culture of the workplace.

   *Organization A*
   - No one in organization A is permitted to dress up casually.
   - Employees revere their superiors and keep away from unnecessary disputes.

- The people are extremely particular about the deadlines and make certain the responsibilities are completed within the specific time frame.

*Organization B*

- The employees can wear whatsoever they feel like.
- People in organization B are least worried about work and spend their utmost time loitering and gossiping around.
- The employees use belittling comments at the work place and pull each other into controversies.

In the above case, employees in organization A wear dresses that radiate professionalism and rigorously follow the guiding principle of the organization. On the contrary, employees in organization B have a laid back approach and do not take their job seriously. Organization A follows a stringent professional culture whereas organization B follows a feeble culture where the employees do not agree to things enthusiastically.

2. **Values**
   The next level according to Schein which makes up an organizational culture is the values of the employees. The values of the people working in the organization play an essential function in deciding the organizational culture. The thought process and approach of employees have a profound impact on the culture of any particular organization. What people think in fact matters a lot for the organization. The frame of mind of the individual connected with any particular organization influences the culture of the place of work.

3. **Assumed Values**
   The third level is the assumed values of the employees which can't be calculated but do make a difference to the culture of the organization. There are firm values and details which stay concealed but do have an effect on the culture of the organization. The inner features of human nature come under the third level of organization culture. The organizations follow certain practices which are not talked about frequently but understood nevertheless. Such rules form the third level of the organization culture.

## *Changing culture*

In the midst of so many components, you can observe why altering culture is such a problem. Culture is a defensive mechanism. All the assumptions work in partnership to support and sustain the other assumptions. Even when you try to change one in isolation, the other assumptions have a tendency to buttress conventional behavior.

The assumptions are also enforced by the individuals or groups who have authority within the organization. If you desire to change the culture, you regularly have to change the leaders, either in their personality or physically change them. The former is frequently so difficult that the latter is the only alternative.

If you are in an executive or leadership position, you need to take a step back and jot down how your organization represents each of the assumptions mentioned above. Just be conscious that your answer will be tainted by your own influence on that culture. This is where the worth of an outside impartial viewpoint can be important. Culture is something we all experience but have trouble defining. Optimistically, this helped present a model to elucidate why we do what we do when we get together for general goals.

An organization's culture according to Schein is determined by the following primary factors [14]:

1. Observed behavior: language, customs, traditions
2. Accepted group norms: standards and values
3. Espoused values: published, publicly announced values.
4. Formal philosophy: mission and vision
5. Rules of the game: rules for all in the organization
6. Climate: climate of group in interaction
7. Embedded skills
8. Habits of thinking, acting, paradigms: shared knowledge for socialization
9. Shared meanings of the group
10. Metaphors or symbols

## *Three elements of culture*

1. **The Problem of Socialization—Teaching Newcomers**
   Socialization in simple terms means to get ready as newcomers to become members of an existing group and to feel, believe, and do something in a way that the group considers suitable. Seen from the group's point of view, it is a procedure of replacing a member. Such extensively diverse situations as child upbringing, teaching somebody a new game, acquainting yourself with a new member of an organization, training somebody who has been in sales work to become a manager, or familiarizing an immigrant with the life and culture of a new society are all examples of socialization.

   Socialization is an essential process in communal life. Its significance has been noted by sociologists for a long time, but their illustration of it has altered over the last hundred years. In the early days of American sociology, socialization was compared with civilization. The concern was one of refining fierce people so that they would freely work together with others on general activities. An uncontrollable human nature was assumed to be present before an individual's encounter with civilization. This personality had to be twisted to do the accepted thing in acceptable ways.

However, as time went on, socialization came to be seen more and more as the end effect, that is, as internalization. Internalization denotes taking social norms, roles, and values into one's own mind. Society was observed as the principal factor in charge of how people discovered how to think and act. This observation is apparent in the work of functionalist Talcott Parsons, who gave no indication that the effect of socialization may be doubtful or may differ from person to person. If people fall short in playing their anticipated roles or behaved weirdly, functionalists explain this in terms of deficient or insufficient socialization. Such people are said to be *unsocialized* meaning they have not hitherto learned what is anticipated of them [15].

a. *Social Position as Part of the Context*:

   Family's societal class, financial position, and cultural background—as well as your gender—can have an effect on the ways in which you will be socialized. People in more advantageous positions tend to develop advanced self-evaluations. As a consequence, they feel vindicated in having more wherewithal. Likewise, those in less-desired positions have a tendency to have inferior self-evaluations and may feel that their lower status is deserved [16].

   Sociologists inquire if kids in diverse societal classes are socialized in different ways. For example, are middle-class kids socialized differently from lower-class kids? If so, why and how? Middle-class parents are to some extent less likely to use bodily punishment than are lower-class parents [17]. Middle-class parents emerge to be more troubled about their children's motives than with the harmful consequences of their actions. Consequently, if a child breaks a dish for instance, a middle-class parent will be anxious as to whether he or she did it "on purpose" or whether it was an accident and the response will differ accordingly. Lower-class parents have a propensity to react in about a similar way no matter what the objective of the child was [18].

   These dissimilarities in parental reactions may arise from the life circumstances of people in different classes. Different parental understanding in the professional world colors the outlook of what children need to learn [18,19].

2. **The Problem of Behavior—Cultural Predisposition and Situational Contingencies Behavior Is Derivative, Not Central**
   This recognized definition of culture does not consist of obvious behavior patterns (even though a number of such behaviors—predominantly prescribed rituals—do echo cultural assumptions). Instead, it lays emphasis on the significant assumptions that deal with how we observe, imagine, and think about things. Obvious behavior is always decided both by the cultural predisposition (the perceptions, feelings, and mind-set that are patterned) and by the situational emergencies that arise from the direct exterior environment.

   Behavioral regularities can happen for causes other than joint culture. For instance, if we detect that all members of a group shrink in the presence of a huge, strident leader, this may possibly be based on biological, spontaneous effect reactions to sound and size, or on individual or shared learning. Such a behavioral reliability should not, for that reason, be the base for defining culture, in spite of the fact that we might later find out that, in a given group's experience, shrinking is in reality a result of shared learning and, as a result, a demonstration of deeper shared assumptions. To put it a different way, when we detect behavior regularities, we do not discern whether or not we are dealing with a cultural manifestation. After the deeper levels that describe the essence of a culture are identified the artifacts that reflect the culture can be determined.

3. **Do Occupations Have Cultures?**
   If an occupation absorbs an extreme period of education and apprenticeship, there will surely be a joint education of values, standards, and principles that ultimately will become taken-for-granted assumptions for the members of those occupations. It is understood that the way of life and values learned during this time will remain stable as assumptions even though the person might not always be in a group of occupational peers. However, strengthening of those assumptions occurs at professional meetings and long-lasting teaching sessions, and by good feature of the reality that the practice of the occupation frequently calls for joint effort among a number of members of the occupation, who strengthen each other. One reason why so many occupations rely a great deal on peer-group assessment is that this process conserves and guards the culture of the occupation.

   To determine which set of assumptions is relevant to a whole society, or an entire organization or a full subgroup within an organization or occupation, a study should be carried out empirically. I have found all kinds of combinations; their existence is one reason why some theorists emphasize that organizational cultures can be integrated, differentiated, or fragmented [20]. Nevertheless, for the reason of describing culture, it is imperative to identify that a disjointed or differentiated organizational culture usually reproduces an assortment of subcultures, and inside those subcultures there are shared assumptions.

## *Analyzing existing organizational structure*

The American and the global place of work has been altered significantly with the combination of technological resources that permit employees to work distantly and in circulated locations around the globe as they all support the achievement of a common goal. For illustration, a senior vice-president of marketing may be located in a different nation many thousand miles away from the team providing support. Consequently, comprehending organizational structure is even more essential, given the nonexistence of the customary arrangements or order and authority that are present in traditional organizations.

Research has revealed that organizations will usually encompass key components that can be differentiated along three basic dimensions. Henry Mintzberg [21,22] suggests that organizations can be differentiated along three basic dimensions:

1. **The Key Parts of the Organization**
   These refer to those parts of an organization that play a major role in determining its success or failure. The key parts of an organization and their respective descriptions are as follows (Figure 3.2):
   a. The *strategic apex* refers to top management and its support staff. In an educational setup, this refers to the overseer of schools and the administrative cabinet.
   b. The *operative core* refers to the workers who in reality carry out the organization's everyday jobs. Teachers make up the operative core in school settings.
   c. The *middle line* comprises the middle and lower-level management. Principals are the middle-level managers in school settings.
   d. The *technostructure* comprises analysts such as researchers, engineers, planners, accountants, and personnel managers. In an educational setup, divisions such as instruction, business, personnel, public relations, research and development, and the like make up the technostructure.
   e. The *support staff* are the people who offer indirect services. In school settings, related services comprise maintenance, clerical service, food service, busing, legal counsel, and consulting to provide support.

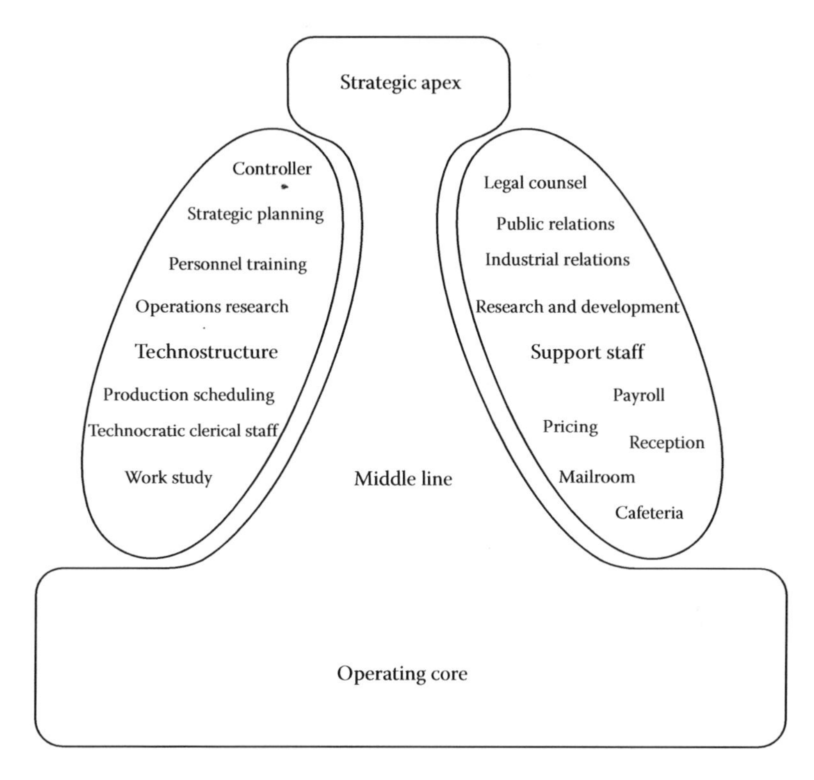


*Figure 3.2* The key parts of an organization.

2. **The Prime Coordinating Mechanism**
   Coordinating mechanism is the main technique an organization uses to coordinate its activities, and it is the second basic dimension of an organization. This includes the following:
   a. *Direct supervision* indicates that one person is accountable for the work of others. This notion refers to the unity of command and scalar principles.
   b. *Standardization of work process* is present when the content of work is specified or programmed. In school settings, this signifies job depictions that oversee the work performance of educators.
   c. *Standardization of skills* is present when the type of training needed to do the work is specified. In school settings, this indicates state certificates necessary for the different occupants of a school settings hierarchy.
   d. *Standardization of output* is present when the outcomes of the work are specified for the reason that the "raw material" that is processed by the operative core (teachers) consists of people (students), and not things, and consistency of output is more

complex to quantify in schools than in other nonservice organizations. However, a movement toward the standardization of output in schools in recent years has come about. Case in point includes competency testing of teachers, state-mandated testing of students, state-mandated curricula, prescriptive learning objectives, and other efforts toward legislated learning.

e. *Mutual adjustment* is present when work is synchronized through unofficial communication. Mutual modification or coordination is the major thrust of Likert's [23] "linking-pin" concept.

3. **The Type of Decentralization Used**
The third basic dimension of an organization is the type of decentralization it employs. This reflects the extent to which the organization involves subordinates in the decision-making process. The three types of decentralization are the following:
   a. *Vertical decentralization* is the distribution of power down the chain of command or shared authority between superordinates and subordinates in any organization.
   b. *Horizontal decentralization* is the extent to which nonadministrators (including staff) make decisions, or it is the shared authority between line and staff.
   c. *Selective decentralization* is the extent to which decision-making power is delegated to different units within the organization. In school districts, these units might include instruction, business, personnel, public relations, and research and development divisions.

Utilizing the three basic dimensions, namely the key parts of the organization, the prime coordinating mechanism, and the type of decentralization, Mintzberg suggests that the strategy an organization adopts and the extent to which it practices that strategy result in five structural configurations: simple structure, machine bureaucracy, professional bureaucracy, divisionalized form, and adhocracy. Table 3.1 summarizes the three basic dimensions associated with each of the five structural configurations.

## *Conclusions*

Henry Mintzberg [21,22] suggests that organizations can be differentiated along three basic dimensions:

1. The key part of the organization, that is, the part of the organization that plays the major role in determining its success or failure.

***Table 3.1*** Structural configuration of organizations according to Mintzberg

| Structural configuration | Prime coordinating mechanism | Key part of organization | Type of decentralization |
|---|---|---|---|
| Simple structure | Direct supervision | Strategic apex | Vertical and horizontal centralization |
| Machine bureaucracy | Standardization of work processes | Technostructure | Limited horizontal decentralization |
| Professional bureaucracy | Standardization of skills | Operating core | Vertical and horizontal decentralization |
| Divisionalized form | Standardization of outputs | Middle line | Limited vertical decentralization |
| Adhocracy | Mutual adjustment | —[a] | Selective decentralization |

[a] In administrative adhocracies the support staff are a key component. In operating adhocracies the operating core is pivotal.

2. The prime coordinating mechanism, that is, the major method the organization uses to coordinate its activities.
3. The type of decentralization used, that is, the extent to which the organization involves subordinates in the decision-making process.

In his 1979 book, *The Structuring of Organizations*, Mintzberg acknowledged five types of "ideal" organizational structures. The categorization was extended 10 years later in the book *Mintzberg on Management* and the following more comprehensive analysis of organization types were drawn up [24]:

- **The entrepreneurial organization:** Small staff, loose division of labor, little management hierarchy, informal, with power focused on the chief executive.
- **The machine organization:** Highly specialized, routine operating tasks, formal communication, large operating units, tasks grouped under functions, elaborate administrative systems, central decision-making, and a sharp distinction between line and staff.
- **The diversified organization:** A set of semi-autonomous units under a central administrative structure. The units are usually called divisions and the central administration is referred to as the headquarters.

- **The professional organization:** Commonly found in hospitals, universities, public agencies, and a firm doing routine work; this structure relies on the skills and knowledge of professional staff in order to function. All such organizations "produce standardized products or services."
- **The innovative organization:** This is what Mintzberg sees as the modern organization: one that is flexible, rejecting any form of bureaucracy, and avoiding emphasis on planning and control systems. Innovation is achieved by hiring experts, giving them power, training and developing them, and employing them in multidiscipline teams that work in an atmosphere unbounded by conventional specialisms and differentiation.
- **The missionary organization:** It is the mission that comes above everything else in such organizations, and the mission is clear, focused, distinctive, and inspiring. Staff readily identify with the mission, share common values, and are motivated by their own zeal and enthusiasm.

Possibly the most distinguishing attribute of Mintzberg's research results and writing on business strategy is that they have regularly placed emphasis on the significance of developing strategy, which arises unceremoniously at any level in an organization, as a substitute or a complement to premeditated strategy, which is determined deliberately either by top management or with the consent of top management. He has been strongly critical of the torrent of strategy literature which centers principally on premeditated strategy.

## *Total systems thinking*

Comprehending the correlation between leadership, innovation, organizational structure, and organizational culture can make the difference in individual occupation accomplishment as well as organizational achievement. First and foremost, consider how these relationships impact an organization by reviewing the research discussed in the study, "Linking organizational culture, structure, strategy, and organizational effectiveness: Mediating role of knowledge management." In this study, the authors performed an analysis of 301 organizations to assess the relationship between the related aspects of an organization.

Practices of knowledge management are context-specific and they can influence organizational efficiency. The study examines the likely mediating role of knowledge management in the relationship between organizational culture, structure, strategy, and organizational effectiveness. A survey was conducted involving 301 organizations. The results suggest that knowledge management completely mediates the impact of organizational culture on organizational efficacy, and in part mediates the impact of organizational structure and strategy on organizational effectiveness. The findings carry theoretical implications for knowledge management literature as they expand the scope of research on knowledge management from probing a set of independent management practices to examining a system-wide mechanism that unite internal resources and competitive advantage.

The study results shed light on numerous unsettled issues in the literature. Apart from providing experimental proof of the correlation between knowledge management and organizational efficiency, this study advocates that knowledge management could be an overriding mechanism between organizational context and organizational effectiveness. The outcome supports the knowledge-based observation of the firm in that knowledge management is not just an autonomous managerial practice, but in addition a central mechanism that leverages organizational, cultural, structural, and strategic influence on organizational effectiveness. It as well matches up with the view of Penrose [25] that the effectiveness of organizational resources varies with alterations in organizational knowledge. Knowledge management serves as a key leverage point in organizations.

Next, organizational strategy applies a noteworthy impact on organizational efficacy above and beyond that of organizational context, even though its result is lessened when organizational culture and structure are taken into consideration. It as well has a momentous impact on knowledge management. These findings demand further examination of the strategy's relationship with knowledge management.

Furthermore, this study makes available some insights in incorporating the resource-based view and knowledge-based view. It discloses that the resources in an organization may be hierarchical. Knowledge may be one step closer to organizational efficiency in the paths leading from organizational resources to organizational effectiveness. Additional investigation is required to study this suggestion.

Lastly, knowledge management was discovered to completely mediate organizational culture's influence on organizational efficacy. This finding advocates that how well knowledge is managed is basically connected with how well cultural principles are translated into significance to the organization. In addition, culture has a superior contribution to knowledge management than other dynamics examined. This may be appropriate to the fact that culture decides the fundamental beliefs, standards, and norms concerning the why and how of knowledge generation, distribution, and use in an organization. This finding reinforces the call for attention to creating an organizational culture that is favorable to learning and knowledge management [26–28].

Several existing studies have focused on the direct relationship between organizational culture and organizational effectiveness. In the current study, however, it has been shown that organizational culture's influence on organizational effectiveness is negligible when a mediator (in this case, knowledge management) is measured. The outcome of this study sheds light on the insufficiency of examining just the direct linkage between organizational culture and organizational effectiveness. It seems that a rational next step in research on culture and effectiveness could press onto a deeper level by investigating the precise mechanism(s) through which organizational culture sways organizational performance.

Even though this study presents considerable answers to some unanswered issues in literature, the outcome should be interpreted in light of its limitations. A key drawback is that the respondents were typically the only informants from their organizations. Only 36 companies of the 301 companies had multiple respondents (12 percent). The single informants may not embody the reality of their organizations as well as multiple informants because single informants may over-report or underreport certain phenomena [29].

## *Managerial implications*

A lot of organizations still see knowledge management as introducing some software programs without sufficient contemplation of their organizational uniqueness to make certain the achievement of their knowledge management programs. Through analyzing the relevance of organizational characteristics to knowledge management success, this study brings to attention the importance of focusing on building a knowledge-friendly atmosphere that is made up of suitable cultural, structural, and strategic elements.

The findings of the study indicate that knowledge management can influence organizational efficacy when it is in coalition with organizational culture, structure, and strategy. Attention on knowledge management practices, such as providing knowledge management tools, and sustaining knowledge management programs, would help transfer the effect of organizational resources to the bottom line.

Next, among the three organizational factors, culture has the most powerful encouraging influence on knowledge management. This entails that knowledge management practices need to center on incorporating culture building activities to promote an environment that is knowledge-friendly. The four dimensions of organizational culture—adaptability, consistency, involvement, and mission—when united optimistically add to knowledge management. They can offer knowledge management professionals a blueprint with regard to which parts of organizational culture to devote their efforts to improve knowledge management results.

## The global leadership and organizational behavior effectiveness project

The effect of culture on individual leadership efficacy is additionally established on a global stage in the result of the global study known as the GLOBE (Global Leadership and Organizational Behavior Effectiveness) Research Project. At the time of publication, the GLOBE study was the most all-inclusive study to date that empirically studied the relationship between culture and leader behavior in such a variety of cultures, with reflection of a variety of quantitative and qualitative procedures as well as having a span of diverse organizations appraised in the analysis (Figure 3.3).

The GLOBE Research Project is a global group of social scientists and management scholars who studied cross-cultural leadership.

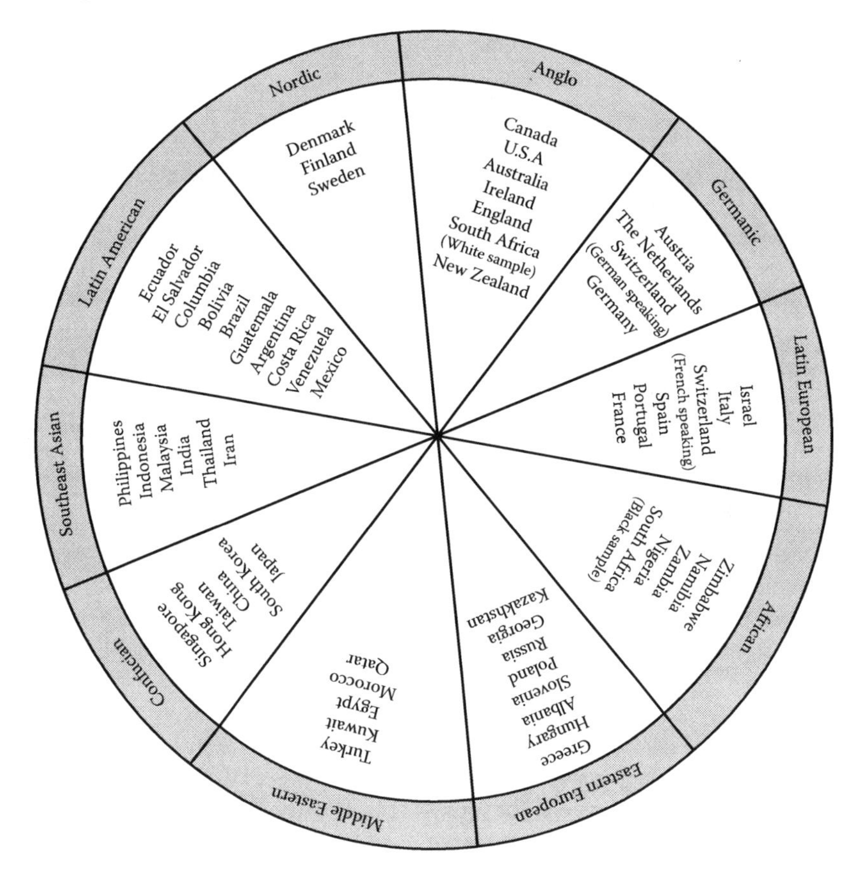


*Figure 3.3* Country clusters according to GLOBE. (Adapted from House, R. J. et al. 2007. *Culture and leadership across the world: The GLOBE book of in-depth studies of 25 societies.*)

Under the GLOBE Research Project, a global group of social scientists and management scholars studied cross-cultural leadership. In 1993, Robert J. House founded the project at the University of Pennsylvania. The project studied 62 societies with different cultures, which were studied by researchers working in their home countries. This global team collected data from 17,300 middle managers in 951 organizations. They used qualitative methods to aid their development of quantitative instruments. The research identified 9 cultural competencies and grouped the 62 countries into 10 geographic clusters, including Latin American, Nordic European, Sub-Saharan, and Confucian Asian.

### *Bases for leadership comparisons*

The GLOBE Research Project identified nine cultural dimensions, referred to as competencies, with which the leadership approaches within geographic group can be evaluated and contrasted:

1. *Performance orientation* refers to the degree to which an organization or society supports and recompenses group members for performance enhancement and brilliance.
2. *Assertiveness orientation* is the level to which persons in organizations or societies are assertive, challenging, and hostile in social relationships.
3. *Future orientation* is the level to which persons in organizations or societies employ future-oriented behaviors such as planning, investing in the future, and postponing pleasure.
4. *Human orientation* is the extent to which persons in organizations or societies support and reward individuals for being reasonable, selfless, pleasant, charitable, kind, and helpful to others.
5. *Collectivism I* (institutional collectivism) is the extent to which organizational and societal institutional practices promote and reward collective sharing of resources and collective deed.
6. *Collectivism II* (in-group collectivism) is the extent to which persons convey pride, devotion, and cohesiveness in their organizations or families.
7. *Gender egalitarianism* is the level to which an organization or a society diminishes gender position differences and gender prejudice.
8. *Power distance* is the extent to which members of an organization or society anticipate and consent that authority should be unequally shared.
9. *Uncertainty avoidance* is the degree to which members of an organization or society struggle to steer clear of uncertainty by dependence on social norms, rituals, and routine practices to ease the unpredictability of future actions.

## *GLOBE leadership dimensions*

After a wide-ranging review of the study, GLOBE contributors grouped leadership features into six dimensions. Researchers then made suggestions about how dimensions of culture and leadership could differentiate behavior in one nation or culture from another.

Identified as the six GLOBE dimensions of culturally approved inherent leadership, these leadership dimensions comprise:

1. **Charismatic or Value Based**
   Portrayed by reliability and decisiveness; performance-oriented by appearing imaginative, motivating, and self-sacrificing; can in addition be noxious and allow for tyrannical command.
2. **Team Oriented**
   Portrayed by negotiation, managerial competence, team partnership, and integration.
3. **Self-Protective**
   Characterized by self-centeredness, face-saving, and procedural behavior capable of inducing conflict when necessary, while being conscious of status.
4. **Participative**
   Portrayed by nonautocratic behavior that promotes participation and engagement and that is encouraging of those who are being led.
5. **Human Orientation**
   Portrayed by humility and consideration for others in a humane manner.
6. **Autonomous**
   Portrayed by capacity to function without regular consultation.

## *Cultural dimensions and culture clusters*

GLOBE's main principle (and finding) is that leader efficacy is contextual, that is, it is entrenched in the social and organizational norms, values, and beliefs of the people being led. In other words, to be perceived as efficient, the time-tested axiom continues to apply: "When in Rome, do as the Romans do."

As a first step to measure leader efficiency across cultures, GLOBE empirically created nine cultural dimensions that make it possible to capture the likeness and/or dissimilarities in norms, values, beliefs, and practices among societies. They developed on the findings by

Hofstede [30], Schwartz [31], Norman and Smith [32], Inglehart [33], and others. They are:

- **Power distance:** The degree to which members of a collective expect power to be distributed equally.
- **Uncertainty avoidance:** The extent to which a society, organization, or group relies on social norms, rules, and procedures to alleviate unpredictability of future events.
- **Humane orientation:** The degree to which a collective encourages and rewards individuals for being fair, altruistic, generous, caring, and kind to others.
- **Collectivism I:** (Institutional) The degree to which organizational and societal institutional practices encourage and reward collective distribution of resources and collective action.
- **Collectivism II:** (In-Group) The degree to which individuals express pride, loyalty, and cohesiveness in their organizations or families.
- **Assertiveness:** The degree to which individuals are assertive, confrontational, and aggressive in their relationships with others.
- **Gender egalitarianism:** The degree to which a collective minimizes gender inequality.
- **Future orientation:** The extent to which individuals engage in future-oriented behaviors such as delaying gratification, planning, and investing in the future.
- **Performance orientation:** The degree to which a collective encourages and rewards group members for performance improvement and excellence.

In conclusion, understanding and managing the relationship between leadership, innovation, and organizational culture can be very useful in building a successful career.

## *Summary*

To become a successful technical leader or manager in an organization needs application of knowledge of the overall organization to one's leadership and management practices. All businesses have an organizational culture no matter how large or small. A business can unofficially build up a culture without the guiding hand of management or ownership, or the company can produce its own culture using a system of values and performance standards. A manager's role in a company's culture is dependent on how the business needs the manager to interrelate with other employees and how much authority the business gives the manager.

Despite the organizational culture, a manager must serve as the representation for that culture for other employees to imitate. For instance, a small-business owner craving to see more employee collaboration must have a manager who is capable of working openly with employees and promoting a team atmosphere. In view of the fact that the culture of a business may swing over time, this will necessitate a manager to become flexible and effortlessly adjustable to change. The faster a manager can demonstrate the appropriate model of a company's preferred culture, the sooner employees will accept it.

A manager's responsibility inside the culture of a small business might necessitate him or her to reward workers who appropriately exhibit the company's preferred characteristics. Rewards can take the shape of simple praise inside the place of work or may comprise higher grades on performance evaluation, which can lead to promotions and superior rates of pay. Rewarding workers for attaining suitable organizational culture demonstrates to the staff that owners and management value every worker's place in the company and are serious about upholding standards.

Leaders shape the manner people reason and conduct themselves because leaders are seen by others as role models, and workers look around to see if their attitudes are consistent with the organization's advocated principles and beliefs. Leaders set the agenda. Leaders sway the organization's culture and in turn the long-term efficiency of the organization. Leaders and managers set the framework within which organizational members struggle for distinction and work together to accomplish organizational goals.

Several different researches evidently draw attention to the relationship between leadership strategies, the impact leaders have on others, and efficiency in the leadership function. This impact has massive importance in assisting to comprehend organizational culture and the role that norms and expectations play in organizational efficiency.

Leadership helps in shaping culture. Culture in turn helps to shape leadership. Together they both compel performance.

---

**Case study: Entrepreneurship (boldness/courage pursuing career opportunities)**

There are many engineers who have successfully made the transition from engineer to entrepreneur in order to lead some of the world's greatest companies. Take, for instance, Larry Page (Google), Bill Gates (Microsoft), or Larry Ellison (Oracle).

However, despite these illustrious examples, most entrepreneurs are *not* engineers, and while not mutually exclusive, engineering and entrepreneurship often require different skills. In order to successfully transition from engineer to entrepreneur, engineers often lean on the expertise of entrepreneurs as well as engineers who have successfully taken products to market.

Deepak Chopra, renowned physician and author of several *New York Times* best sellers, founded the Chopra Foundation to study the scientific effects of mind–body practices on health and well-being. As a scientist, this venture was outside of his comfort zone. His advice to aspiring scientists-turned-entrepreneurs as a result of this experience is to embrace uncertainty. Chopra recently wrote in a LinkedIn post: "There is wisdom in uncertainty—it opens a door to the unknown, and only from the unknown can life be renewed constantly" [34]. Getting comfortable with uncertainty can be difficult for engineers, who have been educated in a field where precision is greatly emphasized. But entrepreneurship involves taking risks, and the uncertainty that comes with doing something nobody has ever done before.

Travis Kalanick, CEO and cofounder of Uber, recently told students at the Indian Institute of Technology in Bombay that, "Engineers are the best founders." His advice? Being an entrepreneur takes guts. He told students, "Lots of people have good ideas. Uber is a good idea. I don't know if we were the first people to think about this idea, but we were the first ones to do it. So you've got to just get out there and do it. You have to have enough guts to go out there and try" [35].

This guidance, while somewhat intangible, is essential to entrepreneurship. Having the guts to take risks is not something you can necessarily be taught. In this way, entrepreneurship not only takes intellect but also a great deal of courage and the willingness to fail before you succeed.

Kalanick, who studied computer engineering at UCLA before dropping out to start Scour Inc. (a multimedia search engine which ultimately had to declare bankruptcy), credits a lot of his success to his failure. Pursuing opportunities often involve some degree of failure, but those failures become valuable learning experiences that can contribute to later success. However, to those wondering when it's time to give up on a failing endeavor, Kalanick advises:

"When you are talking about, 'I will lose my sanity for real', that's when it's time to move on," referencing his decision to sell a failing venture after four years of working without a salary [36]. The bottom line? Learning to succeed means learning to fail, and learning to fail means learning when it's time to quit. Each of these lessons comes with experiences that will shape who you are and test your leadership skills. For that reason, you should embrace and look forward to such experiences as you begin pursuing career opportunities in engineering and entrepreneurship.

**QUESTIONS**

1. What engineering leadership characteristics were demonstrated in this case study?
2. How were these characteristics applied to deliver the intended impact?
3. What was the technical, societal, or environmental impact due to the application of the leadership principle(s) identified above?

---

**Real profiles in engineering leadership**

**Name:** Douglas Baltz

**Current position or field of expertise:** Albert Einstein Distinguished Educator Fellow assigned to the National Science Foundation—Division of Undergraduate Education. Analyzing the Noyce Teacher portfolio, STEM Research & Design, STEM Leadership/Policy, and Strategic Educational Partnerships.

**What I like most about my position:** I enjoy the professional expertise and collegial relationships with the NSF personnel. Also, my voice as a current science practitioner is welcomed and sometimes expected during STEM education forums. I like having exposure to federal educational initiatives and current issues in STEM education.

**Who or what has made a difference in my career:** Dr. Pamela McCauley has been an inspirational mentor and educational colleague. She has provided STEM leadership strategies/tools that have guided my career path and, more importantly, encouragement to embrace productive struggle. I am passionate about educational best practices and helping teachers develop their craft. As a result, being a teacher learning lab facilitator and an adjunct instructor at Oakland University's Teacher Development and Educational Studies has had a major difference in my career.

**I felt like quitting when:** When education reform is at a crossroads and the path chosen is not recommended by experts in the field. It is very frustrating when educators are not given a voice at the table when decisions affect ALL students. I feel like quitting when my expertise is not leveraged for the better of ALL students.

**My strategies for success are:** Listening with a purpose. Have a sincere curiosity. Embrace productive struggle and perseverance. Balance humility and humor. Navigate ideas into action. Take a risk and think outside the silo. STEM really stands for "Strategies That Engage Minds."

**I am excited to be working on:** Strategic partnerships in STEM education. Envisioning current and future pathways of educational collaborations is intriguing. This entails analyzing how partners secure a long-term investment (not just a philanthropy check) which brings partners together to assess strategic inputs and design criteria based on standards (NGSS), demographic equity, best practices, educational research data, and global economic workforce needs. Many of the educational partnerships fall into support categories for a strategic purpose. The needs/protocols which describe active educational partnerships include the following categories: Training/Professional Development/Mentoring, Financial Support/Fundraising, Publicity/Recruitment, Program Management/Administration, Internships/Field Experiences, and Data Collection/Evaluation.

**My life is:** My life centers around my loving and supportive family. My career focuses on providing educational opportunities for ALL students to be successful.

I have been an AP Physics and STEM teacher for more than 20 years at Seaholm High School in Birmingham, Michigan. I am serving my Albert Einstein Fellowship at the National Science Foundation, Education and Human Resources Directorate, Division of Undergraduate Education. The AEF comes on the heels of the PASCO STEM Educator of the Year award. This is a national award sponsored by the National Science Teachers' Association and includes nominees of STEM colleagues throughout the nation.

My colleagues regard me as a STEM outreach specialist—building talent pipelines and partnerships with professionals and surrounding higher education institutions. I am the director of Physics Explorer and Operation STEM Summer Camps, and I supervise students effectively in the STEM × research and design course where students have opportunities to experience authentic scientific data collection.

I have presented at numerous National Science Conferences, moderated the panel discussions for the 2015 NASA STEM Education Summit, presented a peer-reviewed paper titled; STEM Research and Design: "A Mentoring Data Experience" at the University of New Mexico, and was a keynote speaker at the 2016 Michigan Science Education Leadership Association Symposium. Title of presentation: *Strategic Partnerships: Envisioning Current and Future Pathways in STEM Education.*

---

## References

1. Zaccaro, S. J. & Klimoski, R. J. The nature of organizational leadership: An introduction. In S. J. Zaccaro & R. J. Klimoski (Eds.), *The nature of organizational leadership* (pp. 3–41). San Francisco, CA: Jossey-Bass, 2002.
2. Fleishman, E. A., et al. Taxonomic efforts in the description of leader behavior: A synthesis and functional interpretation. *Leadership Quarterly,* 2, 245–287, 1991.
3. Hackman, J. R., Walton, R. E. & Goodman, P. S. Leading groups in organizations. In P. S. Goodman et al. (Eds.), *Designing effective work groups* (pp. 72–119). San Francisco, CA: Jossey-Bass, 1986.
4. Mumford, M. D. Leadership in the organizational context: Conceptual approach and its application. *Journal of Applied Social Psychology,* 16, 212–226, 1986.
5. Katz, D. & Kahn, R. L. *The social psychology of organizations* (2nd ed.). New York: Wiley, 1978.
6. Jacobs, T. O. & Jaques, E. Military executive leadership. In K. E. Clark & M. B. Clark (Eds.), *Measures of leadership* (pp. 281–295). West Orange, NJ: Leadership Library of America, 1990, xvii, 636 pp.
7. Thomas, J. B., et al. Strategic sense making and organizational performance: Linkages among scanning, interpretation, action, and outcomes. *Academy of Management Journal,* 36, 239–270, 1993.
8. Huff, A. *Mapping strategic thought.* New York: Wiley, 1990.
9. Jacobs, T. O. & Jaques, E. Leadership in complex systems. In J. Zeidner (Ed.), *Human productivity enhancement* (pp. 7–65). New York: Praeger, 1987.
10. Jacobs, O. T. & Lewis, P. Leadership requirements in stratified systems. In R. L. Phillips & J. G. Hunt (Eds.), *Strategic leadership: A multiorganizational-level perspective.* Westport, CT: Quorum Books, 1992.
11. Zaccaro, S. J. The contingency model and executive leadership. In F. J. Ammarino & F. Dansereau (Eds.), *Leadership: The multiple level approach* (pp. 125–134). Greenwich, CT: JAI Press, 1998.
12. Mumford, M. D. *Cognitive and temperament predictors of executive ability: Principles for developing leadership capacity.* Alexandria, VA: U.S. Army Research Institute for the Behavioral & Social Sciences, 1993.
13. Schein, E. H. *Organizational culture and leadership.* The Jossey-Bass Business & Management Series. Kindle Edition. Kindle Locations 589–592. Wiley, 2010.
14. Schein, E. H. *Organizational culture and leadership/Edgar H. Schein.* 3rd ed. p. cm. The Jossey-Bass Business & Management Series. A Wiley Imprint, CA, 2004.

15. Parsons, T. The present status of 'structural-functional' theory in sociology. In T. Parsons (Ed.), *Social systems and the evolution of action theory* (pp. 67–83). New York: The Free Press, 1975.
16. Della Fave, L. R. The meek shall not inherit the earth: Self-evaluation and the legitimacy of stratification. *American Sociological Review,* 45, 955–971, 1980.
17. Gecas, V. The influence of social class on socialization. In W. R. Burr et al. (Eds.), *Contemporary theory theories about the family, research-based theories* (pp. 365–404). Vol. 1. New York: The Press, 1979.
18. Kohn, M. L. *Class and conformity: A study in values.* Homewood, IL: Dorsey Press, 1969 (second edition, University of Chicago Press, 1977).
19. Kohn, M. L. *Work and personality: An inquiry into the impact of social stratification.* International Society of Political Psychology. doi: 10.2307/3791262.
20. Martin, J. *Organizational culture: Mapping the terrain.* Thousand Oaks, CA: Sage, 2002.
21. Mintzberg, H. *Structure in fives: Designing effective organizations.* Upper Saddle River, NJ: Prentice Hall, 1992.
22. Mintzberg, H. *Tracking strategies: Toward a general theory of strategy formation.* New York: Oxford University Press, 2009.
23. Likert, R. *The human organization.* New York: McGraw-Hill, 1967.
24. Mintzberg, H. *The structuring of organizations.* Englewood Cliffs, NJ: Prentice-Hall, 1979.
25. Penrose, E. *The theory of the growth of the firm.* New York: Wiley, 1959.
26. Davenport, T. H. & Prusak, L. *Working knowledge: How organizations manage what they know.* Boston, MA: Harvard Business School Press, 1998.
27. De Long, D. W. & Fahey, L. Diagnosing cultural barriers to knowledge management. *Academy of Management Executive,* 14(4), 113–127, 2000.
28. Watkins, K. E. & Marsick, V. J. *In action: Creating the learning organization.* Alexandria, VA: American Society for Training and Development, 1996.
29. Gold, A. H., et al. Knowledge management: An organizational capabilities perspective. *Journal of Management Information Systems,* 18(1), 185–214, 2001.
30. Hofstede, G. *Culture's consequences.* Beverly Hills, CA: Sage, 1980.
31. Schwartz, S. H. Beyond individualism/collectivism: New cultural dimensions of values. In U. Kim et al. (Eds.), *Individualism and collectivism: Theory, methods, and applications* (pp. 85–119). Thousand Oaks, CA: Sage, 1994, xix, 338 pp.
32. Norman, P. & Smith, L. The theory of planned behavior and exercise: An investigation into the role of prior behavior, behavioral intention, and attitude variability. *European Journal of Social Psychology,* 12, 403–415, 1995.
33. Inglehart, R. *Modernization and post-modernization: Cultural, economic, and political change in 43 societies.* Princeton, NJ: Princeton University Press, 1997.
34. Chopra, D. *If I were 22: The Wisdom of Uncertainty.* 2014. Available at: https://www.linkedin.com/pulse/20140519230044-75054000-if-i-were-22-the-wisdom-of-uncertainty, accessed on November 17, 2016.
35. CEO.com. *Engineers Are the Best Founders: Uber's CEO Tells IIT Students.* 2016. Available at: http://www.ceo.com/ceos_in_the_news/engineers-are-the-best-founders-ubers-ceo-tells-iit-students/, accessed on November 17, 2016.
36. Business Insider. *Uber CEO Travis Kalanick: This Is When You Know It's Time To Quit.* 2015. Available at: http://www.businessinsider.com/uber-ceotravis-kalanick-advice-on-when-to-quit-your-startup-2015-10, accessed on November 17, 2016.

# *chapter four*

# *Distinguishing yourself as an engineering leader and learning your engineering leadership style*

Studies show that only 22 percent of Americans trust business leaders. Think about that for a moment—we trust less than one-quarter of our business leaders [1]. This is a real issue in terms of credibility and it directly affects consumers' willingness to invest in and buy a company's products or services. The trust level in politics is similarly low and is continuing to diminish. According to a study by the Pew Research Institute, public trust of Americans in politicians in Washington has declined from 73 percent in 1958 to 24 percent in 2014 [2].

According to a study by the Pew Research Center in 2013, the public trust in the U.S. federal government is at an all-time low. Just 19 percent of Americans said they trust the government in Washington to do what is right "just about always" (3 percent) or "a good number of the time" (16 percent) [2].

In the area of trust, less than three-in-ten Americans conveyed trust in the federal government in all key nationwide polls conducted since July 2007 which represented the period of lowest trust in government in more than 50 years. When the American National Election Study initially asked this question in 1958, 73 percent said they could trust the government just about always or a good number of the time [2].

The attrition of public trust in government started in the 1960s. The percentage of the populace that could trust the federal government to do the right thing almost always or a good number of the time reached an all-time high of 77 percent in 1964. But inside a period of 10 years—an era that included the Vietnam War, civil unrest, and the Watergate scandal—the public trust had plunged by more than half, to 36 percent. Near the last part of the 1970s, merely around a quarter of Americans believed that they could trust the government at least most of the time (Figure 4.1).

In the 1980s, trust in government bounced back before declining in the early to mid-1990s. Nevertheless, as the economy prospered in the late 1990s, assurance in government improved. Also in 2001, the 9/11 terror attacks in the United States altered public attitudes on a number of issues including trust in government.

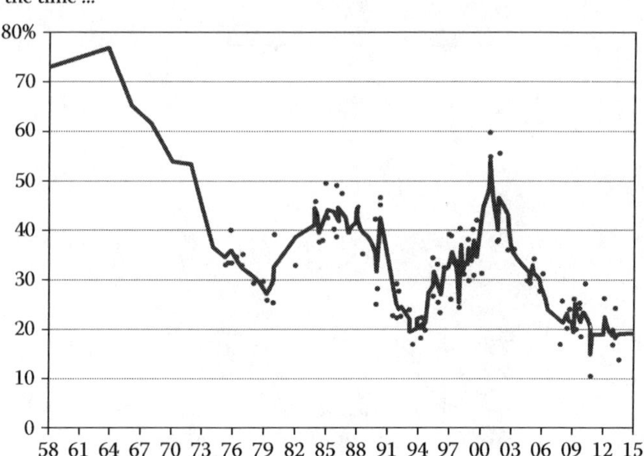


***Figure 4.1*** Public trust in government (1958–2015). From 1976 to 2014, the trend line represents a three-survey moving average. (Survey conducted August 27–October 4, 2015. Q15. Trend sources: Pew Research Center, National Election Studies, Gallup, ABC/Washington Post, CBS/New York Times, and CNN Polls. From 1976–2010 the trend line represents a three-survey moving average. For party analysis, selected datasets obtained from searches of the iPOLL Databank provided by Roper Center for Public Opinion Research, University of Connecticut. Available at: http://www.ropercenter.uconn.edu/.)

A month after the attacks, during early October 2001, 60 percent respondents to the Public Trust survey said they could trust the government, almost twice the number earlier that year and the maximum percentage declaring trust in government in more than 40 years. However, this increase in trust in government was brief—by the summer of 2002, the percentage that said they could trust the government had dropped to 22 percentage points. In the midst of the war in Iraq and economic insecurity at home, trust in government unrelentingly decreased. By July 2007, trust in government had fallen to 24 percent. Ever since then, the percentage that says they can trust the federal government has usually oscillated in a thin range, between 20 percent and 25 percent [2].

Similarly, in the United Kingdom, the trust level for leaders is at a historic low where a study revealed that only 13 percent of people expected the politicians to tell the truth.

A survey poll conducted by Ipsos MORI on December 19, 2014, revealed that the British public is less prone to trust politicians to tell the truth than estate agents, bankers and journalists. Merely 16 percent of Britons trust politicians to tell the truth compared with 22 percent trusting journalists and estate agents and 31 percent who trust bankers [3].

Since 1983, this question has been asked over and over again, which makes it the longest-running series on trust in key careers in the United Kingdom. This assist in drawing attention to the fact that low trust in politicians is long-standing: just 18 percent trusted them to tell the truth in 1983, and in the wake of the expenses scandal, they attained a low peak of only 13 percent trusting them in 2009.

Ipsos MORI is one of the foremost political, social, and business research companies in the United Kingdom. They generate an enormous amount of surveys and research, working with hundreds of clients across the public and private sectors. Their polls comprise tracking data from their research on an extensive array of subject matter, which include education, healthcare, crime, the monarchy, race, business, and politics. Their survey data summarize the observations, experiences, and attitudes of the general public and particular audiences (Figures 4.2 and 4.3).

Other major findings include:

- Doctors remain clearly the most trusted professionals, with 90 percent trusting them to tell the truth.
- Other key public service professions are also highly trusted, including teachers (86 percent), the police (66 percent), and civil servants (55 percent).
- Civil servants in particular have seen a large increase in trust since 1983: only 25 percent said they trusted civil servants to tell the truth in 1983 compared with 55 percent now.

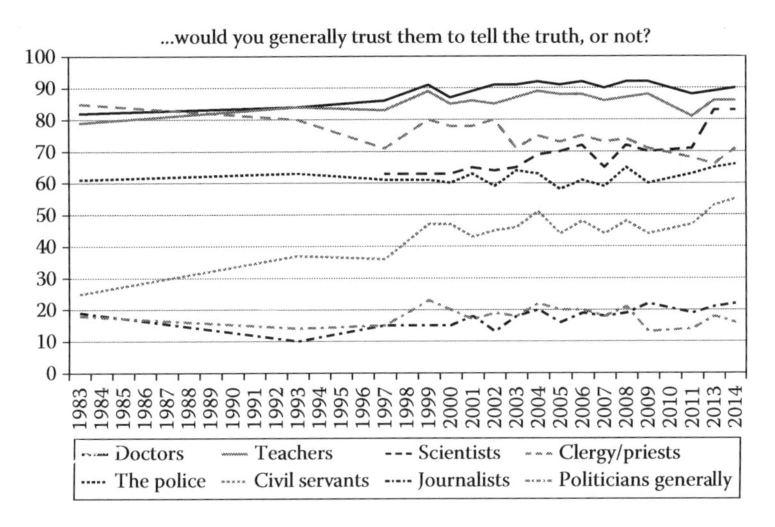


***Figure 4.2*** Ipsos MORI. (From *Social Research Institute*. Survey conducted June 27, 2011. Available at https://www.ipsos-mori.com/researchpublications/researcharchive/2818/Doctors-are-most-trusted-profession-politicians-least-trusted.aspx. With permission.)

# Ipsos MORI

Technical Details

Ipsos MORI interviewed a representative sample of 1116 adults aged 15+ across Great Britain. Interviews were conducted by telephone 5–19 December 2014. Data are weighted to match the profile of the population. Where percentages do not sum to 100 this may be due to computer rounding.

**Trust**

Q1-18 **Now I will read you a list of different types of people. For each, please tell me if you would generally trust them to tell the truth or not?**

*Base: 1166 British adults 15+*

| | Trust to tell the truth | Do not trust to tell the truth | Don't know | Change in trust since last asked |
|---|---|---|---|---|
| | % | % | % | ±% |
| Doctors | 90 | 8 | 2 | +1 |
| Teachers | 86 | 11 | 3 | 0 |
| Scientists | 83 | 14 | 4 | 0 |
| Judges | 80 | 15 | 5 | –2 |
| Clergy/priests | 71 | 24 | 5 | +5 |
| Television news readers | 67 | 27 | 6 | –2 |
| Police | 66 | 29 | 4 | +1 |
| The ordinary man/ woman in the street | 62 | 30 | 8 | –2 |
| Civil Servants | 55 | 38 | 7 | +2 |
| Pollsters | 51 | 37 | 11 | +1 |
| Managers in the NHS | 49 | 43 | 7 | +9* |
| Trade union officials | 39 | 51 | 9 | –2 |
| Business leaders | 32 | 62 | 6 | –2 |
| Bankers | 31 | 65 | 4 | +10 |
| Journalists | 22 | 72 | 6 | +1 |
| Estate agents | 22 | 73 | 5 | –2 |
| Government Ministers | 19 | 76 | 5 | +2* |
| Politicians generally | 16 | 80 | 4 | –2 |

* Last asked in 2011, all others last asked in 2013

***Figure 4.3*** Ipsos MORI Veracity Index 2015. (Adapted from *Trust in Professions.* Available at https://www.ipsos-mori.com/researchpublications/researcharchive/2818/Doctors-are-most-trusted-profession-politicians-least-trusted.aspx.)

- There has also been a consistent increase in trust in scientists in recent years: now 83 percent trust scientists to tell the truth, compared with 63 percent in 1997.
- In contrast, trust in the clergy/priests has declined significantly, from 85 percent in 1983 to 71 percent now.
- Trust in bankers has bounced back from 2013, when only 21 percent trusted them to tell the truth to 31 percent saying they trust them now.
- Managers in the NHS have also seen an increase in trust, from 40 percent when it was last asked in 2011 to 49 percent now [3].

This long-running study however revealed that trust levels are not permanent, and they do swing as the perspective changes. This is evident predominantly in the growing trust in scientists and civil servants and declining trust in the clergy.

There is no way around it; there is a leadership crisis in America and even in the United Kingdom caused by an absence of authentic leaders. Engineers, however, can be part of the solution by understanding and acting on the good characteristics of leaders.

## *Characteristics and attributes of leadership*

Numerous studies have attempted to answer the question: What does it take to be a good leader? Of these studies, a few common characteristics have emerged that consistently surface as key to effective leadership (Figure 4.4). Each of these characteristics is listed below:

1. **Vision**
   Leaders have the gift of seeing ahead. Excellent project managers are capable of foreseeing and tackling challenges that can endanger deadlines, budgets, and user acceptability. An effectual leader is frequently portrayed as someone who has a vision of where to go and the capacity to communicate it. Initially explained by Daniel Goleman in 2002 [4], the visionary leadership style was one of the six leadership types documented. Goleman held that certain leadership styles were more efficient under different conditions. This is identified as situational leadership. For instance, visionary leaders are effectual when a new company is trying to make a move into the marketplace with a product that is very different than products presently obtainable in that marketplace.

   The leader is inspirational in vision and helps others to perceive how they can add to the vision, allowing the leader and followers to move together toward a shared vision of the future. Visionary leaders flourish on change and ability to draw fresh borders, and

| | |
|---|---|
| **Consulting**<br>Checking with others before making plans or decisions that affect them | **Networking**<br>Developing and maintaining relationships with others who may provide information or support resources |
| **Delegating**<br>Authorizing others to have substantial responsibility and discretion | **Planning**<br>Designing objectives, strategies, and procedures for accomplishing goals and coordinating with other parts of the organization in the most efficient manner |
| **Influencing Upward**<br>Affecting others in positions of higher rank | **Problem-Solving**<br>Identifying, analyzing, and acting decisively to remove impediments to work performance |
| **Inspiring Others**<br>Motivating others towards greater enthusiasm for and commitment to work by appealing to emotion, values, logic, and personal example | **Rewarding**<br>Providing praise, recognition, and financial remuneration when appropriate |
| **Intellectually Stimulating**<br>Exciting the abilities of others lo learn, perceive, understand, or reason | **Role Modeling/Setting the Example**<br>Serving as a pattern standard of excellence to be imitated |
| **Mentoring**<br>Facilitating the skill development and career advancement of subordinates | **Supporting**<br>Encouraging, assisting, and providing resources for others |
| **Monitoring**<br>Evaluating the performance of subordinates and the organizational unit for progress and quality | **Team-Building**<br>Encouraging positive identification with the organization unit, encouraging cooperation and constructive conflict resolution |

***Figure 4.4*** Leadership behaviors. (Reprinted from *Transforming Your STEM Career through Leadership and Innovation: Inspiration and Strategies for Women*, Bush, P. M., Copyright 2012, with permission from Elsevier.)

they make it feasible for people to believe they have a real stake in the task. They allow people to experience the vision on their own. According to Bennis, "They offer people opportunities to create their own vision, to explore what the vision will mean to their jobs and lives, and to envision their future as part of the vision for the organization" [5].

According to Dr. Pamela, "The leader knows where she's going because she is led by a vision; this vision, once shared, becomes the guiding force for a team, organization, or corporation. Additionally, a leader must inspire hope and faith—in the vision and oneself. These can only come from a clear and focused vision that we can articulate, accept, and share with others" [6].

Visionary leadership is what every leader ought to have. Whenever there isn't a vision for a team, in that case there is no likelihood of setting a solid goal that everybody can work toward. All leaders should be devoted to that vision, have the skills to identify what the vision should be, and then construe daily tasks and requirements toward that general vision. This can be accomplished via innovation, enhanced efficiencies, and support for team members on a regular basis.

One of the greatest case in point of this type of leadership style came from Steve Jobs. Steve Jobs was once quoted as saying that the difference between a leader and a follower was innovation. He demonstrated this with the improvement of the original iPhone. According to adviseamerica.com [7], as Steve was using the product to test it out, he became conscious that the screen on the original phone test models wasn't just strong enough. It was produced from a plastic resin and it had a tendency to scratch quickly. Jobs looked around for an appropriate replacement, couldn't locate one, but subsequently recalled about a sturdy glass product referred to as Gorilla Glass.

But there was a problem; the company was not manufacturing the product any longer. Jobs called up Corning to get an order placed for the iPhone improvements. At first he was told that it was not doable. However, Jobs knew that Corning could adapt one of their manufacturing amenities to make the glass again within 6 months and so that's what he told them to do. He wouldn't take "no" for an answer and certainly enough, 6 months afterward the iPhone had a glass front. Now several other Smartphones use this glass as well and everybody has found more business success.

2. **Strength of Character**

   In the words of Abraham Lincoln, "Character is like a tree and reputation like a shadow. The shadow is what we think of it; the tree is the real thing." The mental, moral, and spiritual value distinctive to an individual is referred to as character. Character is the sum of one's nature and personality. Your character is inherent in you and it requires character to become a great leader. You can grow your own character strengths, leaders can assist followers grow their character, and organizations can and should enable character development to take place.

   When it comes to leadership, competencies decide what you can do. Obligation decides what you want to do, and character decides what you will do.

   Character is initially for effective decision-making. Obviously, mistakes are made because of a leader's limitation in his or her competencies. More frequently, the core cause is a weakening of character. For instance, not being acquainted with or not being prepared to acknowledge that you don't have the necessary competencies to succeed in the leadership position is entrenched in character. Not keen to listen to those who can do well for the reason that the perception would weaken your leadership is a problem entrenched in character. Defying decisions being made by others but which you feel is wrong need character. Dealing with discriminatory behaviors by others requires character. Creating a culture of beneficial dispute so that others might challenge your decisions without fear of consequences requires character.

Character is such an essential, imperative part of leadership—predominantly for the type of cross-enterprise leadership that is crucial in complex, international business organizations. Character essentially determines how you take on the world around you, what you notice, what you emphasize, who you engage in conversation, what you value, what you choose to act on, and how you decide, etc.

A good example of strength of character is Mahatma Gandhi [8] who was the leader of the Indian independence movement in British-ruled India. He led India to independence and helped to form movements for civil rights and freedom by being an active citizen in nonviolent disobedience. His work has been applied globally for its universality. He displayed strength of character in being an active citizen, socially responsible, loyal, fair, and a team member.

3. **Belief in the Idea, Organization, or Mission**

As a leader, it's imperative that you believe in your organization even as your followers and supporters need to believe in you. You must believe in the mission and vision of the organization. Your belief and convictions in your organization is what will give you the ability to overcome challenges and difficulties that is surely bound to happen in the course of work. A leader must be convinced and committed to the ideas, ethics, goals, projects, and culture of the organization.

It will be disastrous for you as a leader not to be passionate about the organization. Others will never be passionate about the organization if you as the leader is indifferent. It is your passion as a leader that is supposed to influence others and make them want to give their all. It will be very difficult for you to convince others to follow you when you are not passionate about it yourself. Your team is prone to sense hesitation on your part, and in several cases this can lead them to mediocrity, which results in lower than maximum effort and unenthusiastic support for the organization.

Belief is imperative, but blind belief, without regard for others, is risky. Keep in mind that there are several others working toward transformation and development, particularly when these accomplishments can mostly impact the world around you.

For example, assume that you are working toward a more meticulous procedure for product evaluation and introduction whereas another group within the organization is trying to introduce procedures to get original products to the market quicker. These two objectives support the organizational vision but can clash at some level. In such circumstances, don't give up on your convictions but re-examine your conditions, the viewpoint of others, and the atmosphere as it relates to the mission.

4. **Ability to Empower Others**

> Leaders become great not because of their power but, because of their ability to empower others. [16]

Empowering people is one of the major keys to building a high-performance team. Once you empower people by discovering how to stimulate and encourage them, they will desire to work with you and be of assistance to you in achieving your goals in all that you do. Your capacity to develop this positive energy that empowers and inspires individuals will allow you to leverage this passion and achieve considerably more in shorter time frames as a motivated team.

Empowering others come about as you grow into an improved leader and associate. After empowering yourself, the key is to channel this empowerment in addressing challenges that achieve the goals of the organization, impact society and transform lives of others. This means putting together the right combination of team members that have the capacity to successfully address world changing initiatives such as health care challenges, food shortages and energy needs facing the global society. As an engineering leader, you have the capacity to address these challenges given effective leadership skills that will create, inspire and sustain a team in accomplishing these objectives.

There are different types of people that you want to and need to empower on a regular basis. They are the people closest to you: your family, your friends, your spouse, and your children. Next are your work contacts: your staff, your coworkers, your peers, your contemporaries, and even your boss. Third are all the other people that you interact in your everyday life. In each case, your capacity to inspire people and encourage them to support a vision is key to becoming an effective and compassionate leader.

The initial step in empowering people is to abstain from doing anything that disempower them or decrease their power and zest for what they are doing.

There are things you can do every single day to empower people and make them feel good about themselves. Below are a few things leaders can perform to create an atmosphere that empowers people.

- Give authority to those who have established the ability to handle the task.
- Create a positive environment in which people are confident to develop their skills.
- Don't second-guess others' decisions and ideas except when it's really needed. This only weakens their self-confidence and prevents them from sharing potential ideas with you.
- Empower individuals and give them the authority over the necessary resources and individuals to achieve their responsibilities.

Successful leaders are keen to implement their control in such a way that people are empowered to take decisions, share ideas, and attempt new things. A good number of employees perceive the importance of being empowered and are eager to take on the tasks that come with it.

As a leader, your strength is the strength of your team, and, as an engaged leader, you can infuse this strength or power into your team. Nevertheless, this can't be achieved from the outside. To perform this, we have to become skilled at how to be in two places at the same time—in front and in the middle. In front, for the reason that that's where we stand as leaders; in the middle, since that is the heart of our team and the "team" is how we get things accomplished!

Keep in mind that, even though they are a team working jointly for the same goal, the team comprises different individuals, all of whom have their different traits, capacity, and passion. Get to know your team members personally. Allow them to relate to you on a heart-to-heart basis; do not be detached on the hilltop, watching others while they work. Obviously, if you have a big team, getting to be familiar with each person personally will not be feasible. Empower them to be a symbol of the vision and grow your team through them by concentrating on the goal. The fact that you support others to take up leadership responsibilities will improve the self-esteem of your team and inspire them to do more for the cause, aware that their hard work will be acknowledged.

5. **Communication Skills**

As Mike Myatt noted, it's impossible to become a great business leader without being a great communicator—not a big talker, but a great communicator—as well [9].

Renowned entrepreneurs are recognized for their skilled communication with workers, vendors, financiers, and customers. It is one of the most important traits they should possess. Whether the information is positive or negative, they know it is best to be frank, truthful, and timely. They identify that people are pleased about transparency and truth.

The ability to commune with people at every level is just about always identified as the second most essential ability by task managers and team members. Project leadership calls for comprehensible communication regarding goals, tasks, performance, outlook, and feedback.

An immense deal of importance is placed on honesty and sincerity. The project leader is in addition the team's connection to the larger organization. The leader should have the knack of efficiently bargaining and using influence when needed to make certain the team and the project are successful. In the course of efficient communication, project leaders enable individuals and teams succeed by creating a clear plan for achieving results and for the career development of team members. These are demonstrated by communicating concepts, opinions, and ideas in an understandable and concise method, in both verbal and written forms, while interacting with colleagues, administrators, team members, project stakeholders, and others.

All leaders should be good communicators. Successful project leaders efficiently use e-mail, status reports, and meetings to communicate their ideas, get decisions implemented, and resolve challenges. They in addition comprehend that they have to talk about their project in the perspective of whatsoever is most essential to their listeners.

Reflect on the prevailing example of President Abraham Lincoln as he delivered an historic two-minute speech in 1863 at Gettysburg, Pennsylvania. In the middle of the Civil War, he stood where lots of people had died to sanctify their final resting places and to encourage all Americans to keep on the fight for the survival of representative democratic system, that the "government of the people, by the people, for the people, shall not perish from the earth." In this pivotal season of American history, President Lincoln reminded his fellow citizens of the values on which the Founding Fathers established the nation. Those values included human equality and the overarching need to protect the Union. In a few critical moments, with unwavering confidence, passion and knowledge, this leader of the nation spoke plainly about the country's present dark condition yet looked ahead to the bright future. He invited all to stand together in the great cause of humanity. Despite challenges that would follow, due to this exemplary display of leadership, the people received his message and the nation moved together into the bright future which President Lincoln so eloquently described [10].

In conclusion, please keep in mind that every first-rate entrepreneur has and displays the vital skills of communication. It is a key feature of successful business builders.

6. **Ability to Stand Up for Yourself**
   It's not simple being a leader these days. You are in charge of recruiting, training, instructing, monitoring, modeling, envisioning, engaging, inspiring, budgeting, anticipating, prioritizing, building alliances, planning, evaluating, clarifying, adapting, disciplining, directing, reinforcing, recognizing, and reporting.

   For a number of people, a leadership role is the way to authority, a way to bully detractors and coddle followers. Others see it as a means to a trouble-free life, with days spent merging worksheets and sharing out communiqués. However, brilliant workers revolt against the former and pay no attention to the latter. They want to make huge things take place and move forward in their jobs. They push and create, ignore and flout, question and create. This group wants to labor for leaders, not managers. Moreover, they ask the question that managers dread most: *Why?*

   It is just as imperative to turn into a leader as it is to stay one amid all adversities and criticisms. Despite the fact that we are leaders for the cause and not ourselves, as the representative for the cause we have got to be prepared to stand up for ourselves [11]. All of us have at some time faced the barbs, the criticisms, and the attempts to "cut us down to size" [12]. As a leader, it is even more imperative that you become skilled at standing up straight, and fighting for yourself and the cause or the group you are leading. Standing up for yourself means addressing situations, individuals, or rules that limit your movement and the advancement of your cause. It may also mean competing to make sure your cause or issue is seen as a priority in the organization.

   If you are ever to develop into a leader, others will enthusiastically follow; you must be translucent to others and be recognized as somebody who stands by your principles. In addition, as every prospective leader has discerned, foremost you must pay attention to your inner self in order to ascertain who you actually are and what you are all about. There is no scarcity of diverse interests out there contending for your time, your notice, and your endorsement. Prior to you listening to those voices, you have to pay attention to that voice within that tells you what's actually important. Only then will you discern at what time to say yes and at what time to say no—and mean it.

7. **Ability to Make Tough Decisions**
   Everybody makes hundreds of decisions on a daily basis, and as a leader the consequences of your decisions can affect thousands

of people. A lot of our decisions may perhaps have unintentional negative penalty for some, but in search of the dream these choices may be essential.

One of the descriptions of an excellent leader is somebody who can take hard-hitting decisions, somebody who will recognize what has to be done, and is ready to achieve the task-at-hand. An excellent leader can go through with the job despite the fact that it may be horrible or painful. A good leader will have the time to comprehend the rationale, put together a decision, and then move on. Nonetheless, nothing is more perplexing than a leader who is capable of taking tough decisions but instead exhibits an indecisive, indifferent attitude.

There are often time when leaders may be reluctant to make difficult decisions, particularly when they have the potential for a significant impact on individuals or the organization. However, it is imperative to be able to make sound and decisive decisions, even in challenging situations. Being a successful leader requires an ability to address situations where difficult, complex and high impact decision making is necessary. This should be a key aspect of one's professional development strategy as gaining the knowledge, confidence and a process to effectively make challenging decisions is essential to becoming a great leader. You must be capable of making those tough decisions in every aspect of your life, both in business and in personal life.

An example of a good leader is one that is conscious that there possibly will be a need to make some downsizing in the organization. They furthermore recognize the effect of not acting upon the necessity to execute something to guarantee the continuity of the business. They subsequently follow through and "execute the deed."

8. **Willingness to Think Outside the Box**

Thinking outside the box (also thinking out of the box or thinking beyond the box) is a metaphor that means to think differently, unconventionally, or from a new perspective [13]. This expression frequently refers to new or innovative thinking. Out-of-the box thinking entails honesty to innovative ways of seeing the world and an enthusiasm to discover. It involves taking into consideration other ground-breaking options while identifying that new ideas require development and sustenance. Even great innovative people can turn into in-the-box thinkers once they stop trying. As leaders, we need to continuously nurture our capacity to think outside the box and promote this skill in others.

It is imperative not to be restricted to the present but to look toward the future. Given the connectivity of the global society, organizations are deciding that they need to offer a level of authenticity as well as an ability to develop and adapt to changing needs of their customers, clients or constituents. Thus, the leadership must be sensitive to the dynamics of organization and relevant stakeholders. To do this requires formal and informal professional development of those responsible for leading the organization. In other words, to be an industry leader in the global market requires preparation and development of in an organizations leadership team. Great leaders look further than existing practices and markets.

Great leaders think outside the box by exhibiting the following qualities:

a. Listening to others
b. Supporting others when they come up with new ideas and respecting them
c. Valuing new ideas and not being afraid to act on them
d. Willingness to look at new perspectives for day-to-day work
e. Openness to different things
f. Openness to doing things differently

Example of thinking outside the box:

Galileo Galilei, an Italian astronomer, physicist, engineer, philosopher, and mathematician, is a brilliant example of a person who questioned the status quo, "The sun revolves around the earth" was the prevalent thought at the time he lived. Galileo was one of the most renowned outside of the box thinkers [14]. He did not agree to the prevailing theory and what he almost certainly had been taught in school. Instead he opposed the status quo and as a result initiated foundational scientific knowledge that was proven to be a fact. As a result, Galileo Galilei is regarded as a major influence in the scientific revolution during the Renaissance [15].

9. **Patience and Resilience**

The biographies of many great innovators are filled with stories of many people telling them that their ideas cannot materialize. In most cases, they are told to forget their dreams because it was impossible to achieve. Many innovators failed numerous times when trying to accomplish their dream. A good case in point is Thomas Edison in his quest to produce the electric light. But the key to their success was rising up to try again every time they failed.

Nowadays, so many entrepreneurs and leaders want instant success without being prepared to pay the price for success. While success is welcome and desired by all, "overnight" success is a far cry from the actual world of organizational leadership, indicating an approach of patience as a team helps to develop new approaches to tackle challenges and unanticipated situations. This benefits a leader as this is capable of creating an ambience of self-belief, stability, and an air of conviction in the team.

Lots of entrepreneurs and innovators were not able to achieve their dreams for the simple reason that they gave up very quickly. As a leader with an obsessive mental picture, a familiar point of view is to become accustomed, fine-tune, and appraise, but on no account give up: it is always excessively too soon to give up.

John Maxwell, in his book *The 21 Irrefutable Laws of Leadership* [16], tells the account of the McDonald brothers who established a fast-food restaurant, made a success of it, nevertheless failed when they attempted to enlarge the company. They were excellent restaurateurs and business managers, but they were deficient in the expertise to grow to higher levels. They were capable of franchising only 10 new restaurants. As soon as they took in a man named Ray Kroc as a partner he almost immediately became the leader who shaped the international brand which is today known as McDonald's. That is the differentiation between a manager and a leader. Cultivating the patience to tackle a challenge or overcome a failure as you move in the direction of a goal can be the difference between a successful business enterprise and one that is deserted prematurely.

10. **Strong Team Building—Build Your Own Team**
As a leader, it is important to build your own team. This does not necessarily suggest building from scratch. You may possibly have to take over a management arrangement left by a previous manager. Taking over a team previously led by another person comes with a lot of challenges. You will need to lead people you don't know and who don't know you; you will need to work with each and every one of them to produce your own team. Make it obvious that you are at present their leader; however, do not be egotistical while doing so. Share your ideas with them. Give details on your method of working. Pay attention to what they have to say. Don't disregard grievance and questions. Identify with them and attempt to deal with them compassionately. If they are to become your team they need to work with you, not in ill feeling of you. Allow them recognize that you need them a great deal as much as they need you. Immediately you obtain their allegiance, they become your team. By no means take their devotion for granted. The team culture should express importance of loyal behavior, or else it can disintegrate.

As soon as given a chance, to the extent feasible, build your own team. Selection of team members is an extremely important responsibility for a leader. Team composition should reflect and support the mission of the organization, address the stakeholder requirements and community expectations. Additionally, the procedure to create a team should recognizes and recruit individuals with expertise, diversity, key associations, and the ability to successfully achieve the mission of the organization at the uppermost level. As you assemble your team, take account of your improvement soundboard and core teammates. Have a special accountability partner to support you alongside in the midst of a team to keep you accountable for your deadlines and objectives, keep you on your feet, and assist you where you lack the ability or knowledge for the task required.

A good number of us already have a team without being conscious of it. You almost certainly have an important person you converse with about taking care of your children or a partner with whom you have a discussion on the subject of finance. It is essential that you identify with the important idea that you need to have a team, build a team, discover people presently part of your world that are on your team, and go out to incorporate others who will help you actualize your vision.

11. **Determination and Optimism**

In history, there has never been a leader who has not been overwhelmed by a number of uncertainties and indecision. Acknowledge that nothing in life is certain and that a little quantity of ambiguity is a regular element of life and is indispensable as we embark upon fresh challenges. There will be suspicions, hazards, and the unanticipated. Your accomplishments as a leader depend on how determined you are to carry on in the face of these challenges. Bear in mind that your followers are all observing the leader. If your willpower is in doubt, it will not be long before the resolve of your team begins to diminish.

It is important to mention that being strong-willed does mean being stubborn or narrow-minded. These are negative qualities that will always work against you [17]. Inability to identify when to stop when you are engaged in unwholesome activities is an illustration of stubbornness; on the contrary, declining to pull out from a fight that you discern you're destined to win is called strength of mind or determination. It is prudent to perform a tactical withdrawal on an activity if determined necessary. However with thoughtful planning, a tactical withdrawal can be executed in a manner that still allows achievement of the ultimate goal.

Never be worried about the future. Be a determined optimist. It's a fact that we have a complex road ahead of us, one that will at times call for making tough decisions. Yet, it is better to address the challenges of today while focusing on the potential of tomorrow. Don't allow yourself to get weighed down by pessimism. Just as leaders cannot lead from the rear, they cannot inspire in isolation—and negativity and pessimism are isolating features. So don't permit yourself to get weighed down by pessimism. Embrace the authority of determined optimism sharpened with grit, persistence, and hard work.

## *Global leadership*

Today's engineer has an opportunity like never before to make an impact at the international level given the degree of connectivity in business, industry, and social society. As such, an emerging leader must have a mind-set far beyond one's personal geography and instead have a consistent mind-set that reflects concern for the global society. *Global leadership* has been defined by Cohen (2010:5) as:

> Global Leadership is the ability to influence individuals, groups, organizations, and systems that have different intellectual, social, and psychological knowledge or intelligence from your own. [11]

This global leadership is manifested in a leader's ability to not only think globally but also to act in a way that reflects this thinking: In other words, the ability to "think and act globally." A study by Goldsmith et al. (2003) attempted to further define global leadership by interviewing an age- and gender-diverse group of human resource development (HRD) officers in 200 global organizations [18]. This international collection of HRD professionals were asked about the requisite important leadership skills (with 1 = important and 10 = extremely important from a list of 72 items) for effectiveness in the past, present, and future. Not surprisingly, the top three items mentioned for past and present leaders included none remotely suggesting global considerations. However, when considering needs for future leaders, the third most highly rated item was "makes decisions that reflect global considerations." Conversely, this same item was rated as the 70th and 71st, respectively, for the required leadership skills in the past and the present [11]. So, while global considerations were not considered that important just six years ago, the aggressive globalization of our society has HRD professionals recognizing that a change in mind-set is critical for future leaders.

The study went on to identify global leadership skills categorized into five overall clusters:

1. Thinking globally
2. Appreciating cultural diversity
3. Being tech savvy
4. Building partnerships and alliances
5. Sharing leadership

These clusters of qualities have a synergistic impact and when exercised they will equip an individual to value, connect, and guide international leadership efforts. It is important to note that these five qualities for global leadership are to be used in conjunction with the previously discussed attributes of effective leaders.

A global leadership mind-set involves carefully balancing three overall dichotomies [11] and these dichotomies as quoted from Cohen can be described below:

1. *Global formalization versus local flexibility,* wherein formal approaches unify an organization in the customer's eyes so that they know what to expect from the global brand, but the manifestation of this brand locally may look different. For example, organizations may create different packaging of products more familiar to local customs and expectations while never changing their brand image.
2. *Global standardization versus local customization,* wherein minimally standard protocols and processes are needed to create one company "way," but there needs to be flexibility in how these are implemented at the local level based on local requirements. For example, depending on local regulatory controls, some food and drug specifications may vary accordingly, while never endangering consumers.
3. *Global dictate versus local delegation,* wherein ways of doing business need to be uniform but local implementation must be delegated according to existing customs. For example, local customs might dictate how business is conducted but never violate defined corporate values.

In this discussion, *Fortune* magazine cites McDonald's as an example for managing these polarities on a global stage. The McDonald's business model allows countries to create their own buns, bags, and business practices that cater to local tastes [19], while maintaining strong branding in other areas of the business. Similar principles have been applied in technology. An example of this is in the work schedule and environment of technical employees in many multinational companies. This has led to an increase in telework and virtual teams that are deployed by global companies to do more cost-efficient and time efficient work. This allows companies to consider the local preferences, offer flexible work environments while still attaining the technology development goals of the organization [20].

## The role of emotional intelligence

*Emotional intelligence* (EQ or EI) can be defined as the ability to understand, manage, and effectively express one's own feelings, as well as engage and navigate successfully with those of others. In previous years, there was no strong appreciation for "soft skills" such as emotional intelligence (EI). However, in recent years EI has been increasingly seen as an essential skill for global leaders. In fact, the difference between *good leaders* and *great leaders* has been defined as the level of EI or their EI rating.

EI is not the traditional intelligence quotient (IQ), college degrees, or technical skills. According to Goleman, the components of EI are as follows [17]:

1. ***Self-Awareness*: Knowing One's Strengths, Weaknesses, Drives, Values, and Impact on Others**
   The deep perception or comprehension of one's emotions, strengths, weaknesses, needs, and drives is what is referred to as self-awareness. People that have strong self-awareness are usually honest with themselves and with others. They are neither excessively critical nor idealistically positive. They are acquainted with how their thoughts affect them, other people, and the performance in their job.

   When you have self-awareness, you are prone to be honest, open, and have the capability to assess things and yourself realistically. Self-aware individuals usually tend to be frank, outspoken, and sincere regarding their emotions and the impact it has on their job functions. Individuals with self-awareness can easily be recognized during performance evaluations. They know about their strength and weakness and have no problem talking about them. Self-aware people even appreciate constructive criticisms and are not put off by it. This trait is contrary to people with low self-awareness who take criticism as a threat and indication of failure.

2. ***Self-Regulation*: Controlling or Redirecting Disruptive Impulses and Moods**
   The ability to self-regulate yourself is very vital for a leader. Being able to self-regulate means that you are able to control or redirect impulses and moods together with the inclination to think before acting. Just like every other person, people who are successful in self-regulating themselves will sometimes find themselves in bad moods and emotional impulses; nevertheless, they always look for ways to control those feelings.

   Your ability to self-regulate as a leader is very essential for the reason that your ability to control your feelings and impulses will enable you to create an atmosphere of trust and fairness. Moreover, the ability to self-regulate is very essential to be able to adapt to the constantly fast-moving and ever-changing business environment. Individuals that have the ability to self-regulate are able to move with the constant change in the business environment. They have mastered their emotions and as such they don't panic when a new initiative is announced. They instead are able to seek out new information and ideas that will enable them to adapt.

Self-regulation also develops both personal and organizational trust. A lot of terrible events that take place in companies are a reaction of impulsive behavior. People usually do not of necessity plan to lie, exaggerate, or misrepresent facts; however as soon as a chance presents itself, people with low impulse control immediately say yes. By contrast, individuals who possess high levels of self-regulation are able to challenge and conquer impulses which help them to put up lasting relations built on trust. Leaders who are able to emotionally self-regulate consequently have an inclination for reflection and consideration; are comfortable with uncertainty and change; and have the capability to challenge and overcome impulsive urges.

3. ***Motivation*: Relishing Achievement for Its Own Sake**
Are you a motivated leader? Leaders who are effective have a drive to achieve beyond the expectations of the organization and even their own personal expectations. A good number of people are motivated by external factors such as large salaries, big titles, and whatever it is that they dream about achieving. Those with leadership potentials are nonetheless self-motivated. The desire to achieve, the passion for their work, and the deep cravings to create drive a leader. Leaders have a passion for the work itself, search for innovative challenges, are keen to learn, and take immense pride in a job well done. They are habitually restless with the status quo, seek to do things better, and are ready to discover new ways to perform their work better.

Individuals that are motivated to attain great things are ceaselessly lifting the performance level. Individuals that are motivated to do better in addition desire ways of monitoring both their progress and that of their organization. Individuals that have low achievement motivation are frequently unclear about results.

Individuals with lofty enthusiasm and passion are always positive even when circumstances seem against them. In such situations, a leader combines self-regulation with achievement motivation to prevail over the distress and despair that come following a setback or disappointment. The drive to accomplish results eventually transforms into strong leadership. Leaders that set high performance standards for themselves will likewise do the same for their organization. In the same way, a passion to exceed objectives and an interest to track results can be infectious. Leaders that exhibit these characteristics frequently have the capacity to put together a team with the same characteristics.

4. ***Empathy*: Understanding Other People's Emotional Makeup**
Empathy is the capacity to understand or feel what another being (a human or nonhuman animal) is experiencing from within the

other being's frame of reference, that is, the capacity to place oneself in another's position [21]. Empathy refers to the ability to compassionately consider the feelings of employees in addition with other necessary factors when trying to make intelligent decisions. Empathy is distinct from sympathy, pity, and emotional contagion [22]. Sympathy or empathic concern is the feeling of compassion or concern for another, the wish to see them better off or happier. Pity is feeling that another is in trouble and in need of help as they cannot fix their problems themselves, often described as "feeling sorry" for someone. Emotional contagion is when a person imitatively "catches" the emotions that others are showing without necessarily recognizing this is happening [23].

Considering others' emotional state of mind doesn't connote taking up other people's sentiments as one's own or trying to make everybody satisfied—because that would simply make it impossible to take action. Leaders with empathy identify with the emotional makeup of people and they can distinguish what people are feeling. People with empathy are accustomed to details in body language—they can perceive the message underneath the words being spoken.

Empathy is an essential element of leadership in today's team-based business settings. A leader within teams must be capable of sensing and comprehending the perspectives of everybody in the group and influencing them to speak candidly about their thoughts.

5. ***Social Skill*: Building Rapport with Others to Move them in Desired Directions**

Social skill is any skill facilitating interaction and communication with others. Social rules and relations are created, communicated, and changed in verbal and nonverbal ways. The process of learning these skills is called *socialization* [24].

Social skills are the skills we use to commune and interrelate with each other, both orally and non-orally, through gesticulation, body language, and our personal appearance. Human beings are sociable creatures and we have developed numerous ways to communicate our messages, feelings, and thoughts with others.

Socially skilled people have a tendency to have a wide circle of friends and they find common ground with people of all kinds. It doesn't indicate that they socialize frequently, but it does indicate that they work according to the postulation that nothing significant gets done single-handedly. Such people have a set of connections in place when the moment for action comes.

People who are socially skilled tend to be effective in managing teams. They excel in persuading people to support a vision through a demonstration of self-awareness, self-regulation, and empathy put together. Individuals who are excellent persuaders discern when to make an emotional request and when an appeal of reason will pay off better. Passion makes socially skilled people exceptional partners or group members—their obsession for the work infects others and they are motivated to achieve results.

Sometimes, socially skilled people may seem not to be working while they are actually socializing. They can be chatting in the walkway with contemporaries or chitchatting around with people who are not even linked to their "actual jobs." Socially skilled people don't see any sense in subjectively limiting the span of their relationships. They build connections extensively for the reason that they discern that in these fluid times they might require help sooner or later from people they are presently getting to know today. Social skill is a key leadership competence. Leaders must manage relations efficiently. A leader's job is to get work achieved through other people and social skills make that achievable.

## *How to improve your emotional intelligence*

> When our emotional health is in a bad state, so is our level of self-esteem. We have to slow down and deal with what is troubling us, so that we can enjoy the simple joy of being happy and at peace with ourselves. [25]

According to Talent Smart, 90 percent of high performers at the workplace possess high EQ, while 80 percent of low performers have low EQ. EI is absolutely essential in the formation, development, maintenance, and enhancement of close personal relationships. Unlike IQ, which does not change significantly over a lifetime, our EQ can evolve and increase with our desire to learn and grow [26].

Below are some positive ways to improve your EI:

1. **Keep a Daily Journal**
   Keeping a journal can help you develop your self-awareness. When you jot down your thoughts and feelings daily, you can move yourself to a higher level of self-awareness. Recognizing what you did, why you did it, how you did it, and how it makes you feel, as well as what you did well will help you determine what you need to improve on. Merely reflecting on every day, will develop your self-awareness and personal improvement habit.

2. **Calm**
   Our lives, especially in a demanding work environment, can be chaotic most times. Sometimes, everything just seems like a blur. Learning to always calm down no matter the demanding situation will not only help you to be more productive but also actually help you to have peace of mind in the midst of difficult circumstances.

   One easy way to achieve this is to pause and ask yourself some little questions: Why are you feeling annoyed or agitated? What can be done to counter this? An emotion is a state of mind—how can you switch your anger to a more positive feeling? What can you discover from this condition? No matter what the circumstance, you can always choose to be calm when reacting to it.

3. **Understanding Your Values**
   To improve your EI, you must take out time to recognize what your beliefs and values are, comprehending your motivating values, that is, your passions and drive behind your ambition. Furthermore, list out your principles, passions, and values. It may also be beneficial to list out those feelings that make you draw back; the factors that cause you to avoid things because of fear. Take out some time to identify your highly valued principles as well. These are essential values and principles that you do not want to compromise. When you distinguish what are important to you, decisions will be much easier to take—decisions that do not compromise your values.

4. **Be Accountable for Your Actions**
   Keep in mind you have an alternative for all you do. If you didn't previously, take responsibility for all your actions and decisions from now on. If you make a terrible decision, gain knowledge from it, nevertheless face the music and take responsibility for it. You achieve deference when you do and with deference comes great leadership.

5. **Goals**
   For each one of your goals, write down at least two reasons why you absolutely must accomplish them! These reasons will confer your inspiration to go on when times are tough.

6. **Turn Negative Situations Into Positives Ones**
   Each moment in time when things are hard, inquire yourself, "What can be learned from this?" Is there something that I can take away and initiate from this situation so that this kind of circumstances doesn't occur again? Jot down your learning points in your journal.

7. **Learn and Understand Conflict Resolution**
   As a leader you must learn how to resolve conflicts between your team members, clients, or vendors. Learning these skills is very important if you desire to do well as a leader.

8. **Learn How to Praise Others**
   As a leader, you can instigate the allegiance of your team basically by praising them when they deserve it. Learning how to efficiently praise others is an excellent skill that is well worth.

## *Leadership styles*

From the sources of leadership, we can parse out leadership by focus areas and from the components of those areas where leadership is exercised we will focus on two categories: the task leader and the person leader. Although there is often overlap and a leadership position often involves both, these categories or areas of leadership can be defined as follows:

- A task leader focuses on tasks, concerned by the what, how, where, when, and by whom a task is being accomplished.
- A person leader focuses on the people who do that task, particularly on their welfare and satisfaction, and is concerned with group cohesion, satisfaction, and conflict resolution.

Each of these leadership styles will be influenced by the leadership competencies and the style of the individual responsible for executing the desired leadership need.

### *Six leadership styles*

A combination of EI competencies is needed to complete most tasks. The focus should be on the desired outcome and use the aforementioned in the most appropriate situations.

The most important job of a leader is to achieve results. Nevertheless, even with all the training programs for leadership and numerous acclaimed experts available today, effective leadership still eludes several people and organizations. One cause of this according to Daniel Goleman is that such experts offer advice based on inference, experience, and instinct, and not on quantitative data [4]. Based on research of more than 3000 executives, Goleman explores which specific leadership behaviors produce affirmative results. He delineates six different leadership styles, each one arising from different components of EI. All the approaches have separate consequences on the working atmosphere of an organization, division, or team, and, in turn, on its financial performance. The approaches by name and short explanation alone will reverberate with anybody who leads, is led, or, as is the situation with the majority of us, does both.

1. **Authoritative Leaders**
   The authoritative leader was first described by Daniel Goleman in combination with the six leadership styles described in his presumption of EI [26]. As explained by Goleman, authoritative leaders

are specialists in their field of work. They are individuals who are capable of clearly articulating a vision and a pathway to success. The mark of this kind of leaders is their capacity to organize people in the direction of a vision. This leadership style is most effective when an innovative vision is desirable or at a time the course of that vision is not constantly clear. One fascinating feature of this style is that although the leader is regarded as an authority, he or she nevertheless lets his or her followers to figure out the best way to achieve their goals.

a. **Pros and Cons:** The authoritative leadership style is best suited in circumstances when an organization or teams appear to be drifting without any direction. For instance, it's helpful when a group or organization has been cut off, and their overall approach and purpose within the larger organization is no longer understandable to the followers.

   *Pros*: A study conducted by Hay/McBer [26] looked at the notes of thousands of executives to comprehend their behaviors and their effect on the work environment. The desire of the team was to better comprehend how a specific leadership method affects their direct reports. The results of the study point out that authoritative leadership has the most positive impact of all the different leadership styles on the general working environment. This style produces a very optimistic and cheerful place to work.

   *Cons*: In spite of the fact that authoritative leadership has the most optimistic effect on the work environment, it still isn't automatically the style that should be utilized every time. In particular, if a new manager finds himself or herself positioned in a workgroup of specialists, it may be hard, if not unattainable, for this new leader to come into the group and instantly articulate his or her idea of where the workgroup should go.

**Examples:**

**John F. Kennedy**—One of the several attributes that John F. Kennedy is remembered for today is his vision regarding the space program of the United States. While speaking at Rice University, on September 12, 1962, President Kennedy said:

> We choose to go to the moon in this decade and do the other things, not because they are easy, but because they are hard, because that goal will serve to organize and measure the best of our energies and skills, because that challenge is one that we are willing to accept, one we are unwilling to postpone, and one which we intend to win …. [27]

President Kennedy further went on to talk about "metal alloys" that had not hitherto been manufactured "that are" able of resisting heat and stresses a number of times more than have ever been experienced. He had a dream of safely sending a man to the moon and back. He even gave details on how it was going to be achieved. President John F. Kennedy was demonstrating an authoritative leadership style that mobilized the wherewithal of a whole country toward a distinct goal.

2. **Coaching Leaders**
When an organization is in need of somebody to share their knowledge or when their intellectual capital is weak, they will require the service of a leader with a coaching style of leadership. Leaders that possess the coaching style are very good at helping others to advance their skills, providing career guidance, and building bench strength
   a. **Attributes:** As expressed by Daniel Goleman, the coaching leadership style approach is most aptly summed up by the expression "try this." Coaching leaders are able to bind together career ambitions and individual goals. They assist followers to perceive how everything fits together. Furthermore, because of this capability and interest in assisting others, they are excellent at developing a long-term plan to achieve long-term goals.
   
   Coaching leaders offer ample feedback on performance; nevertheless, they are in addition experts at handing over and giving people assignments that are challenging as well. These types of leaders won't grasp someone's hand through the rough times, although they will enlighten them how to withstand the storm. In summary, coaching leaders are individuals with an indisputable interest in helping others to be successful. They accomplish this by focusing on the growth of others, by means of their intense sense of empathy, and their own self-awareness.
   b. **Pros and Cons:** One of the fascinating findings of Goldman's study was that the coaching leadership style was the least used in the place of work mostly because many managers do not believe they have the time required to assist others. This mindset is regrettable because the investment made in employees often provides abundant returns.
   
   The coaching style is very useful at improving results. Advanced atmosphere and performance are attained for two reasons that go further than the investment in teaching others:
      i. Coaching leaders make available a very positive work atmosphere.
      ii. Workers know precisely what is required of them and they comprehend the overall strategy of the organization.

Conceivably, the major disadvantage of this leadership style is that it takes time and patience. A leader has to make an initial investment in an employee with the optimism that he will reap the rewards in the form of above average performance later on.

**Examples:**

It is difficult to hit upon clear examples of modern-day coaching leaders. The best illustration of this specific style would be those tagged as "celebrated" mentors or those mixed up in well-known mentoring pairings.

The following list of mentoring pairings offers instances of the coaching leadership style at work:

a. Red Holtzman (NBA coach) mentored Phil Jackson (NBA coach)
b. Andrew Carnegie (philanthropist) mentored Charles Schwab (first president of U.S. Steel)
c. Eddy Merckx (five-time Tour de France winner) mentored Lance Armstrong (seven-time Tour de France winner)
d. Robert Patterson (CEO of National Cash Register) mentored Thomas Watson (founder of IBM) [28].

3. **Affiliative Leaders**

The affiliative leader was most accurately described by Daniel Goleman in combination with the six leadership styles in his theory of EI. As explained by Goleman, affiliative leaders can be summed up as individuals that are frequently more responsive to the importance of people than realizing goals. Affiliative leaders take pleasure in their capability to keep employees joyful and produce a pleasant work atmosphere. These leaders endeavor to put together strong relationships with those being led, with the expectation that these relationships will bring about a strong sense of devotion in their followers.

a. **Pros and Cons:** While working for this type of manager might sound promising, this leadership style does have its limits. Mediocre results may quickly follow if this is the only style used. It's advisable to use the affiliative leadership style in conjunction with one that focuses on achieving business results too.

   *Pros*: Research carried out by Hay/McBer recorded the observations of thousands of executives trying to understand their behaviors and their impact on the work environment [26]. This team sought to better comprehend how a particular leadership approach influences their direct reports. The results of the study signified that the affiliative leader has the second most optimistic impact of all styles, just behind the authoritative leadership style. Both leadership styles are very effective in producing a positive and cheerful workplace.

Affiliative leaders in addition offer their followers with ample positive feedback. They are fast to identify the hard work of others and offer rewards for a job well done. They are tremendously successful at patching up things among members of the team.

*Cons*: Similar to the authoritative leadership style, the affiliative style is not a style that a leader will crave to put into practice all the time. While affiliative leaders are great at offering positive feedback and inspiring team members, they are frequently reticent in dealing with under-performing members in a team. In view of the fact that poor performance can go unrestrained in the team, a number of employees might get the feeling that ordinary performance is sufficient. This could lead to a speedy decline in general team performance.

Affiliative leaders are furthermore unsuccessful when the team is faced with multifaceted problems. In fact, for the reason that the leader offers strictly encouraging feedback, they can unintentionally inspire their followers to continue down an erroneous path.

An effective leader needs to practice various leadership styles and recognize when it's suitable to discontinue or begin utilizing a particular style.

**Examples:**

The standard illustration of an affiliative leader, and the one frequently cited by Goleman, is Joe Torre, the ex-manager of the New York Yankees.

Joe Torre was the manager of one of the most brilliant teams in all of baseball. Amid the entire talents in one place, there will surely be lots of self-centered players to deal with, too. In this setup, perchance one of the utmost deeds of a manager is basically holding the team as one, and building a feeling of accord between teammates. This is a proficiency that affiliative leaders specialize in.

Torre quickly identified the contributions of different players and conveyed his appreciation for the results in the win/loss discourse. By examining his various affirmative statements in the media, and how well he treated his players, it's simple to recognize how efficient an affiliative leader can be in the right environment.

4. **Democratic Leaders**

Lewin et al. way back in 1939 [29] used three styles as the base for their leadership model, namely the autocratic model, the democratic model, and the laissez-faire model. In the democratic leadership style, there is a sense of balance in the decision-making process.

Employees, or followers, are permitted to partake in the decision; their views count just as much as the leader's.

Daniel Goleman in addition thought there was sufficient unique distinctiveness found in democratic leadership to incorporate it as one of his six leadership styles. In his model, the most important actions of these individuals were to build compromise through teamwork. The key to this style is communication, which means a leader has to be on the lookout for the viewpoint of others and share his or her viewpoint, too.

Democratic leaders build confidence, value and dedication by getting people to contribute their ideas and key into the organization's strategy. Staff in democratic organizational settings have a tendency to be very reasonable about what can and cannot be accomplished the organization's strategy and allowing decisions to be made the represent the perspective of the majority of the group.

a. **Pros and Cons:** A good number of us would like to believe that the democratic style can be efficiently applied to any group of workers. Nevertheless, when we begin to scratch beneath the surface, the pros and cons of the democratic leadership style becomes obvious.

   *Pros*: In view of the fact that employees or followers have an equal say in the decision-making process, they are more dedicated to the desired result. The atmosphere of mutual respect created by this leadership style often results in more meticulous resolution to problems. This helps to create an ideal environment for joint problem-solving in addition to decision-making. Nevertheless, this process has its shortcomings, too.

   *Cons*: The democratic leader relies on the knowledge of his followers or workers. If the workers are inexperienced, this style is not especially useful. A fair quantity of proficiency is necessary to make excellent decisions.

   The other weakness of the democratic style is that joint effort takes time. When requesting opinions from people, it takes time for them to clarify what they think and for others to identify with what they are saying. The approach is ineffective if the need is urgent, and the leader should change his leadership style. The democratic leadership style is most successful when there is a place of work that has experienced workers, and there is adequate time to develop a comprehensive solution.

**Examples:**

One of the preeminent examples of a democratic leader is Dwight D. Eisenhower. As a military leader, Eisenhower faced the complicated task of getting Allied forces to consent on a common approach. Eisenhower contrived hard to ensure that everybody worked collectively to arrive at a general understanding [30]. This was one of his best accomplishments. It was at this point that the democratic leadership style, and shared efforts, of Eisenhower shone through. The ensuing victory of the Allied forces supported the suitability of the approach under those circumstances.

5. **Pacesetting Leaders**

The pacesetting leader was also first described by Daniel Goleman in conjunction with the six leadership styles defined in his theory of EI. Although there are known situations where this style is effective, it's essential to use it cautiously.

Pacesetting leaders are as well fast to identify individuals that are not meeting up to their expectations. Below performers are requested to step up their performance and if they fail to do so, they are swiftly replaced.

Pacesetting leaders don't give workers a lot of affirmative feedback; they basically don't have the time. On the flip side, they have no difficulty jumping right in and taking over if they believe improvement is too slow.

a. **Pros and Cons:** Undoubtedly one of the pros of the pacesetting leadership style is that they are capable of quickly achieving organizational results. In the short term, there is often a high-energy group with exceptional performance in terms of achieving responsibilities as well as the superiority of the work itself.

   On the negative side, the style has a downbeat effect on the work environment. In reality, only the coercive leadership style has a greater negative consequence on people. Frequently, employees are purely beleaguered by the pace and the strain placed upon them, which results in quickly declining morals. To worsen matters, the alacrity under this leadership style is such that orders to followers may not even be comprehensible. Contemptuously, the leader has no tolerance for those that need to be taught or are not adapting to the new work fast enough.

**Examples:**

Possibly the best illustration of pacesetting leaders will come from the military. In the conditions experienced in that background, it's imperative to formulate swift and critical moves. There is very little or no patience for making mistakes while the stakes are that high.

6. **Coercive Leaders**

   The concept of a coercive leader was first expressed by Daniel Goleman in combination with the six leadership styles defined in his theory of EI. Although managers realize the need to adjust to different styles to changing conditions, the coercive style is one that should be used with caution.

   Goleman uses the subsequent phrase to sum the style of coercive leaders: "Do as I say." The style is most useful when an organization or group is faced with a crisis. This can vary from dealing with unproductive workers to an absolute turnaround for an organization or group.

   For instance, if a department is having problems with workers pursuing dangerous work practices, the leader might employ the coercive style to achieve instant conformity with the organization's security standards. A manager may in addition utilize this style when an organization or unit is not working gainfully owing to extravagant practices.

   a. **Pros and Cons:** One of the benefits of employing the coercive style is that the leader has an immense deal of control over what is taking place in their specific organization. The leader provides directions and demands obedience. This is mostly successful when an organization is in difficult times.

      On the contrary, research has established that this style of leadership has a very harmful impact on the whole work environment. In fact, by its very character, the coercive styles are nonflexible, provide little compensation, and take away from workers every responsibility for their actions—just as long as they are following instructions.

      Despite the fact that some workers in reality take pleasure in being told precisely what to do every day, the majority will find the coercive style excruciating in the long run.

   **Examples:**

   The coercive leadership style is best utilized in circumstances where the organization or followers need an absolute turnaround effort. For example, it is useful during disasters or dealing with underperforming workers, typically as a last resort. Under those situations, the immediate obedience with an order or instruction hastens the road to recovery.

## *Conclusion*

The *Commanding* leaders insist on immediate conformity without input from others. The *Visionary* leader inspires and gets the team on board

| | Commanding | Visionary | Affiliative | Democratic | Pacesetting | Coaching |
|---|---|---|---|---|---|---|
| The leader's modus operandi | Demands immediate compliance | Mobilizes people toward a vision | Creates harmony and builds emotional bonds | Forges consensus through participation | Sets high standards for performance | Develops people for the future |
| The style in a phrase | "Do what I tell you." | "Come with me" | "People come first" | "What do you think?" | "Do as I do, now" | "Try this." |
| Underlying emotional intelligence competencies | Drive to achieve, initiative, self-control | Self-confidence, empathy, change catalyst | Empathy, building relationships, communication | Collaboration, team leadership, communication | Conscientious-ness, drive to achieve, initiative | Developing others, empathy, self-awareness |
| When the style works best | In a crisis, to kick start a turnaround, or with problem employees | When changes require a new vision, or when a clear direction is needed | To heal rifts in a team or to motivate-people during stressful circumstances | To build buy-in or consensus, or to get input from valuable employees | To get quick results from a highly motivated and competent team | To help an employee improve performance or develop long-term strengths |
| Overall impact on climate | Negative | Most strongly positive | Positive | Positive | Negative | Positive |

***Figure 4.5*** The Six Leadership styles. (From Goleman, D., 2000, Leadership that Gets Results, *Harvard Business Review*, 78, March–April, pp. 82–83. With permission.)

by empowering them with the vision. *Affiliative leaders* build emotional bonds and harmony. *Democratic* leaders build compromise through contribution. *Pacesetting* leaders anticipate excellence and self-direction. And *coaching* leaders build up people for the future (Figure 4.5). The research by Goleman indicates that leaders who get the best results don't depend on just one leadership style; they utilize a good number of styles in any specified week. Goleman details the kind of organizational circumstances that each style is best suited for, and he explains how leaders who do not have one or more styles can develop their suite of professional capabilities to be effective leaders. He asserts that leaders can switch between different leadership styles to create powerful outcomes, thereby converting the art of leadership into a science.

## *Assess your leadership style*

The key to being an effectual and long-lasting leader is being able to lead a mix of different people in an array of circumstances. In order to accomplish this, you need an excellent mix of leadership styles. The following quiz will help you discover the leadership styles you are good at and those you may need to build up further. Knowing your leadership style may assist you comprehend why you lead the way you do, whether changing your style will be simple, and what kind of people you need to engage to balance for some areas of weakness.

The following assessment was adapted from Kappa Kappa Psi and Tau Beta Sigma website [31]:

The assessment is designed to assist students in leadership development.

**Assessing your leadership style**

*Note:* This test is designed to help determine your personal leadership style. There is no right or wrong answer. Just choose the answer which seems most like what you would naturally do.

1. When your chapter is meeting, it is most important for you that:
   a. You stay on schedule and get through the material you planned for the group
   b. You make sure that each person has had a voice in the discussion
   c. You let the discussion run its natural course and see what happens
2. If you are leading a discussion and you find one person dominating it, do you:
   a. Invite others to participate in the discussion
   b. Hope that the person will eventually get the hint and stop talking so much
   c. Tell the person that you'd like others to have a chance to participate
3. You arrive late for an important chapter event and discover that the two freshmen in charge have not set the room up properly and are busy in last minute preparations. Do you:
   a. Figure it's too late to do anything and roll with the punches
   b. Pull them aside and tell them what has to be done
   c. Ask them if they can try to improve the room set up
4. Your chapter secretary has consistently forgotten to submit the proper paperwork to your school's administration. Do you:
   a. Find someone else that can do the job
   b. Ask the secretary about what's going on and offer to help
   c. Ask the secretary to try harder the next time
5. You've just asked another member of the chapter to join the leadership team. The best way to get them started in their new role is to:
   a. Make sure they have an opportunity to really get to know the other chapter leaders

b. Let them have enough "adjustment" time to get used to the new role
c. Make sure they understand very clearly what is expected of them

6. The best way to keep the chapter up to date on schedule changes is to:
   a. Let everyone learn about the changes through regular interaction and let them know if anyone has questions to get in touch with you
   b. Send out an email explaining the changes
   c. Ask the chapter president to put it on the meeting agenda
7. You happen to be one of the main student leaders of your school's band. In a leader's meeting, one of the younger leaders questions a decision you have made. Do you:
   a. Try to explain why you arrived at your decision
   b. Ask the person to elaborate on why they question your decision
   c. Ask the person what decision they would have made
8. You discover that a member of you chapter has been openly critical of your leadership. Do you:
   a. Set up a meeting to discuss why this person has been critical
   b. Wait for the other person to bring it up to you directly
   c. Immediately meet with the person and confront him or her on his or her attitude
9. You are the chapter president and you have some strong thoughts on how to lead the chapter. However, your chapter sponsors disagree. Do you:
   a. Allow a little time to go by and see if the sponsors change their mind
   b. Ask the sponsors to suggest other alternatives that will work
   c. Tell your sponsors that you appreciate the other ideas, but you have strong reasons for your decision and that you need to be trusted
10. When solving a problem that affects others, do you:
   a. Present the problem, the solution, and each person's part in the implementation
   b. Discuss the problem and try to get everyone to agree on a common solution
   c. Trust that each person will solve his or her part of the problem that affects them
11. You are in charge of planning a chapter fundraiser. Do you:
   a. Encourage spontaneous meetings to discuss plans

b. Check to see that everyone knows what to do
c. Let people get in touch with you if they have any questions

12. In establishing a committee to plan an event, it is best to:
    a. Allow the committee to function at their own pace
    b. Ask the committee to establish their own timeline after they understand the objectives
    c. Give the committee clearly defined objectives, a timeline, and standards of operating

13. The best way to handle a difference in opinion between two members in your chapter is to:
    a. State the differences of opinion and present a compromised position that both can accept
    b. Encourage the two to meet together and work out their differences
    c. Bring the two people together and help them arrive at a solution

14. In a group of friends, do you:
    a. Try to make sure everyone has been heard
    b. Enjoy listening to the ideas of others
    c. Easily offer your opinion

**Scoring the leadership profile**

| Question | Directive | Consultative | Free Rein |
|---|---|---|---|
| 1 | A_____ | B_____ | C_____ |
| 2 | C_____ | A_____ | B_____ |
| 3 | B_____ | C_____ | A_____ |
| 4 | A_____ | B_____ | C_____ |
| 5 | C_____ | A_____ | B_____ |
| 6 | B_____ | C_____ | A_____ |
| 7 | A_____ | B_____ | C_____ |
| 8 | C_____ | A_____ | B_____ |
| 9 | C_____ | B_____ | A_____ |
| 10 | A_____ | B_____ | C_____ |
| 11 | B_____ | A_____ | C_____ |
| 12 | C_____ | B_____ | A_____ |
| 13 | A_____ | C_____ | B_____ |
| 14 | C_____ | A_____ | B_____ |
| TOTAL | _______ | _______ | _______ |

My natural leadership style is:_______________________

**Directive leadership style**

1. In challenging situations, you feel most comfortable working with clear guidelines.
2. In meetings, you take charge early and become anxious to get down to business.
3. You find it easy to assign tasks, provide schedules, and monitor progress.
4. You may tend to become impatient when subordinates want to prolong a discussion. You tend to be more concerned with getting the job done than you are with meeting interpersonal needs.
5. In situations in which you have complete control, you tend to relax more, assume an easy-going manner, and become more patient and considerate.

**Consultative leadership style**

1. The primary goal is to have good interpersonal relations with others—even at the expense of the goal.
2. You tend to be very sensitive to the individual members of the group and are especially concerned with their feelings.
3. In a meeting you tend to encourage the participation of various members of the group.
4. In high-stress situations, you tend to find it more difficult to reach the goal.
5. You function best in moderate control situations where you are able to deal with interpersonal relations and deal effectively with difficult subordinates.

**Free rein leadership style**

1. In challenging situations, you allow the greatest freedom to your subordinates.
2. You can become overly tolerant of nonproductive members of your team.
3. Your "best" day is one in which you have spent the majority of your time working on projects and administrative functions.
4. You schedule meetings but may tend to have a difficult time bringing the discussion to any definitive conclusion or implementation plan.
5. You tend to function best with subordinates who enjoy working on their own and need little day-to-day supervision from you.

**Comparison of leadership styles**

| Area of Concern | Directive (Control Orientated) | Consultative (Team Approach) | Free Rein (Laissez-Faire) |
|---|---|---|---|
| Who does the planning? | Leader | Leader plus group | Individuals or groups |
| Who does the problem-solving? | Leader | Leader plus group | Individuals or groups |
| Who makes decisions? | Leader | Leader plus group | Individuals or groups |
| What is the direction of communication? | Down | Down, up, and across | Across |
| Where is the responsibility for achievement felt? | Leader | Leader plus group | Not Felt |
| Where does the responsibility actually lie? | Leader | Leader | Leader |
| Leader's confidence level in subordinates | Little to none | High | High |
| Leader's rapport with subordinates | Low | High | Questionable |
| Amount of delegation of authority by leader | None | Lots | Lots |
| Crisis management | Good | Poor | Chaotic |
| Change management | Poor | Good | Ineffective |

*Source:* Kappa Kappa Psi and Tau Beta Sigma National Headquarters. *Assessing your leadership style.* Available at http://kkytbs.org/mwd/Documents/TBS/Assessing%20Your%20Leadership%20Style.doc, accessed on November 17, 2016.

## Behaviors that work for and against leadership selection

The behavioral approach to leadership suggests that people can behave in ways that may lead to their being selected as leaders or distinguish them as leaders. This is similar to the trait approach, except that traits are considered more or less inborn attributes, while behaviors can be learned and refined. The following list shows behaviors that have been identified as being related to being selected or not selected as a leader in groups.

| Contributing Behaviors | Interfering Behaviors |
|---|---|
| • High participation and talking<br>• Comfort and fluency in delivering information<br>• Forceful and energetic in presentation<br>• Does not express strong opinions early in group (can later) | • Low level of participation, involvement, or contribution<br>• Uninformed contribution<br>• Overly directive comments<br>• Offensive language (including sexist and profanity) |

*Continued*

| Contributing Behaviors | Interfering Behaviors |
|---|---|
| • Initiates conversation<br>• Interacts flexibly with others; changes style as needed but not perceived as a chameleon<br>• Promotes identity of group (we, us, our, etc.)<br>• Listens accurately to others' contributions<br>• Demonstrates achievement, goal orientation, and task structure up (can later) | • Stilted, overly formal language<br>• Shows contempt for leadership<br>• Willing to do as told<br>• Presents self too strongly early in the group discussion fanity) |

*Source:* Kappa Kappa Psi and Tau Beta Sigma National Headquarters. *Assessing your leadership style*. Available at http:/kkytbs.org/mwd/Documents/TBS/Assessing%20Your%20Leadership%20Style.doc, accessed on November 17, 2016.

Go back over the above lists and reflect on the differences between them. What are the principles involved? What makes a person desirable or undesirable as a leader based on behavior?

## *Leadership style development*

Ideally, when technology and leadership come together to meet a known need, the opportunity for innovation is created. In order to increase the likelihood for success as an innovator, it is important to understand the skills of innovators. The *Innovator's DNA* by Jeffrey Dyer, Hal Gregersen, and Clayton Christensen provides a resource for characterizing the skills of innovators [32]. The book is the result of a six-year study in which 3000 creative executives were surveyed and 500 individuals interviewed. The end result was five "discovery skills" that distinguish these individuals as successful innovators:

1. **Associating**
The first and foremost discovery skill is "association." Association refers to the capability to effectively unite apparently unconnected questions, problems, or ideas from diverse fields. When it comes to disorderly innovation, you don't have to begin from scratch to produce something innovative. Raising questions and combining facts across different disciplines and businesses are the means to enhancing your ability for relationship, a cognitive ability that lies at the heart of innovation. In this chapter, the authors Jeffrey Dyer, Hal Gregersen, and Clayton Christensen lectured on how to reinforce your capacity for associating by describing how corporations such as Oracle and Starbucks and gadgets such as the BlackBerry are in truth the invention of association. "Defocusing," "Lego thinking,"

and "curiosity boxes" are just a number of the realistic approaches and activities intended to help you grow the associating skills that might show the way to your own breakthrough innovation.

2. **Questioning**
Sometimes we keep our ideas and apprehensions to ourselves, scared to come across as stupid or spiteful. However, when you stifle your innovative thoughts, you limit your prospects for innovation. In this section, authors Jeffrey Dyer, Hal Gregersen, and Clayton Christensen encourage you to direct your inner mind and relive your spirit of inquisitiveness. They enlightened how asking diverse and challenging questions permit you to reframe challenges and discover new and improved answers, and how questioning can be utilized as the ingenious channel for innovation. In addition, with methods like "Question Storming," this chapter presents a question series that guarantees initiation of your curious thinking in a dynamic course. By means of contemplating what might be as a replacement for concentrating on what is, you can liberate yourself from the chains of custom and fall upon something incredibly new [32].

3. **Observing**
Leaders who are result driven frequently turn out to be so frenzied by the strain of managing an organization that they never pause to observe what is going on about them. Nevertheless, disruptive innovators are continuously watching out for tasks that people want to get done or partial solutions that can be the way out to daily problems. The authors in this chapter encourage you to expand your viewpoint by making use of your marginal vision. They demonstrate to you how observing your surroundings can convey great insight. Furthermore, they specified three habits that will make you an excellent observer: 1) keenly watching clients to see what products they "hire" to do what jobs; 2) Learning to search for surprises or irregularity; and 3) discovering chance to observe in new surroundings [32].

4. **Experimenting**
Disregard the Petri dish, the lab coat, and all those dangerous chemicals. Innovators can conduct experiments in all places, on a daily basis, with whatsoever resources they have at hand. In this section, the authors talked about the various forms experimenting is capable of taking, from hands-on tinkering to theoretical deliberations, as well as narrowing them down to three styles for approaching the process yourself. The best element of this discovery skill is the fact that failure is not a choice. Every test conducted produces a reaction

that is capable of leading to a new discovery or being stored for a future innovation. Innovative entrepreneurs frequently participate in a number of forms of "dynamic experimentation," whether it is academic exploration, material tinkering, or commitment in a new ambience. As leaders of innovative enterprises, they make experimentation essential to all they do. Determined experimentation permits leaders the chance to assemble data and widen their fields of view.

With a basic level of effort and thinking, anyone can become an innovator by incorporating these characteristics into everyday life.

Some people practice these skills more than others. Why? It has to do with courage. Through practice, individuals gain the confidence to regularly use these skills. However, a willingness to do what hasn't been done before is what truly rounds out these characteristics into the description of an innovator.

5. **Networking**
   Ability to think outside the box is just one element of disruptive innovation. Gambling outside your social circle is similarly vital. Expanding your private and professional acquaintances will ignite an outpouring of fresh knowledge and new perspectives that can motivate innovative thoughts. In this section, the authors stress on the rewards of networking with group and corporation outside your area of proficiency. Whether at an official thought networking occasion or an informal backyard meeting, innovators are constantly learning from the people around them. Devoting moment in time and energy to discovering and testing ideas through a network of different individuals gives innovators a fundamentally different viewpoint. Distinct from a good number of executives—who network to access wherewithal, to advertise themselves or their business, or to enhance their careers—innovative entrepreneurs go out of their way to meet people with diverse ideas and viewpoints to broaden their own information domains. To this end, they make a deliberate attempt to visit other nations and meet individuals from other walks of life.

## *What it means to be a technical or technology leader*

- **Product Development Leadership**
  It is the creation of better or more effective products, processes, technologies, or ideas that are valued by the markets, governments, and the society. In business and engineering, new product

development (NPD) is the total process of bringing a new product to market. NPD is portrayed in the literature as the transformation of a market opportunity into a product available for sale [33], and it can be substantial (i.e., something physical you can touch) or insubstantial (like a service, experience, or belief).

An excellent understanding of client needs and wants, the competitive situation, and the nature of the market characterize the top required factors for the success of a new product [34].

- **Organizational Leadership**
  Organizational leadership refers to the management staff that typically provide inspiration, objectives, operational oversight, and other administrative services to a business. Effective organizational leadership can lend a hand in prioritizing objectives for subordinates and can offer guidance in the direction of achieving the overall corporate vision.

  Organizational leadership is a twofold, focused management approach that works toward what is paramount for individuals and what is paramount for a group as a whole at the same time. It is in addition an approach and a work ethic that authorize an individual in any position to lead from the top, middle, or bottom of an organization. Organizational innovation can be applied in several forms to create the innovative, value-added, and profitable results anticipated in the business environment (Figure 4.6).

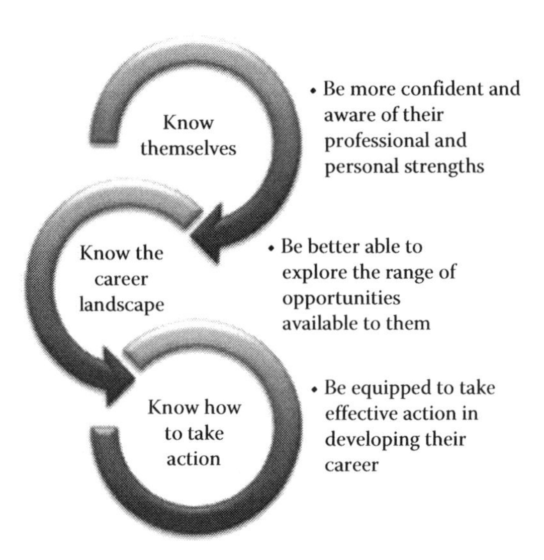


***Figure 4.6*** What innovative leaders need to know. (From Bush, P. M., *Innovative Leadership: Inspiration and Strategies*, Stanford University, October 22, 2013.)

- **Individual Leadership**
  Personal leadership is the ability and desire to crystallize your thinking and to establish a specific direction and destination for your own life. It includes the courage, choice, and commitment to move in that direction by taking committed and determined action to acquire, accomplish, or become whatever you visualize for your future.
  Important to your work and individual success is erecting your own personal leadership. There is no alternate to discovering more about yourself, obtaining facts and insight, practicing new talents, and enhancing more of your individual potential for greatness. Building personal leadership is completely critical to your long-term attainment, accomplishment, and contentment. Nonetheless, several people are so busy that they find no time to build a brighter opportunity for themselves.
  Personal innovation modifies career focus, individual capabilities, skills, and relationships to meet present and emerging needs in a given area of opportunity.

Can you distinguish yourself as an engineering leader? The first step is to understand the characteristics of a good leader/innovator, understand your style, and then apply those characteristics in the areas you need to cultivate. This will help you exercise your leadership as a project manager, in product development or organizational leadership.

---

**Case study: Crisis situations**

In 1998, Bill Gates and Warren Buffet sat down together to share their wisdom about each of their roads to success. Bill Gates said Microsoft's rise to success was not without its bumps in the road, mentioning the rise of the Internet as something that Microsoft really wasn't ready for at first. The Internet was growing much faster than Microsoft had planned for, and as a result they did not have an adequate strategy in place to keep up. According to Gates:

> That was a case where as an act of leadership I had to create a sense of crisis and have a couple months where we all just threw ideas and sent electronic mail around, went on a bunch of retreats ... and then, it eventually coalesced around a few ideas .... [35]

Gates says this kind of crisis can be expected to arise every so often and a good entrepreneur is prepared to chart a new course when necessary in order to keep up with industry trends. Microsoft had to be flexible and adaptive, which can seem counterintuitive from an engineering standpoint. Engineering tends to be about calculated, measurable, and precise goals; however, there are times that call for putting down the script temporarily or, in some cases, re-writing it altogether.

In March of 2015, the Association of Science—Technology Centers (ASTC) asked scientists from around the world to explain how they had handled a variety of crises that tested their leadership and the lessons they learned.* May M. Pagsinohin, executive director of the Philippine Foundation for Science and Technology, has been in charge during several major natural disasters, like typhoons and earthquakes, and from her experience of picking up the pieces each time, she said:

> I feel honored to respond to these unique circumstances where leadership is tested. These disasters have provided the opportunity to move through different stages of difficulty, transformation, recovery, and new beginnings. Drawing from these experiences, I have some lessons to share [36]:
>
> - Never waste a good crisis as it provides a platform to get things done creatively and rise above the challenges. There is no place for "on-the-job training." Don't be caught off guard; always be prepared.
> - Empower your team. Encourage people to develop their confidence and ask them to participate actively.
> - Trust the board. Our open and candid communication with our board and stakeholders helps us review and evaluate our organization's capabilities in the areas of disaster assessment, crisis planning, and organizational recovery.

Whether your crisis is a natural disaster or an unplanned-for industry trend that needs to be addressed, good leadership

* Bimonthly magazine of the Association of Science-Technology Centers Leadership at Every Level Dimensions, March–April 2015, p. 32.

will guide the organization through these important tests and determine whether or not the organization will emerge stronger as a result. Crises are opportunities, and the best leaders are able to remain calm under pressure and can truly shine in these situations.

**QUESTIONS**

1. What engineering leadership characteristics were demonstrated in this case study?
2. How were these characteristics applied to deliver the intended impact?
3. What was the technical, societal, or environmental impact due to the application of the leadership principle(s) identified above?

---

**Real profiles in engineering leadership**

**Name:** Christopher T. Jones

**Current position or field of expertise:** Corporate Vice President and President, Northrop Grumman Technology Services

**What I like most about my position:** I am fortunate to lead a team comprising 14,000 dedicated, talented men and women located throughout the United States, and in many other countries globally. I'm energized by the opportunity to visit these people, understand what they do, and then determine what I can do to make them more effective and successful.

**Who or what has made a difference in my career:** Luckily for me, I was able to determine early in life that I wanted to be involved in the aerospace engineering field, and that I wanted to serve in the U.S. military. My family was fully supportive, which was critical. Since that time, I have had the continuous support of my family, friends, teachers, military professionals, and industry colleagues—too many to list or single out.

**I felt like quitting when:** It is true that there are good days and bad days in any job or activity. However, I do think about our dedicated employees who are working hard and are depending on me to support them. I often reflect, too, on the men and women throughout the world who are reliant on the equipment, service, and capability that my team provides. Many of them are deployed or directly involved in operations—their situations are far more stressful than anything I am facing now.

**My strategies for success are:** For a long time, I have used an approach that involves three priorities. First, I try to provide those whom I lead with whatever they need to be successful. Next, I try to provide my teammates, partners, and superiors with what they need to be successful. Finally, if there is any time left, then I worry about myself and what I need. As I have progressed in my life, I find that I am spending more time on #1 and #2, and much less on #3.

**I am excited to be working on:** There are lots of new technologies and innovative concepts being developed by Northrop Grumman, and by lots of other people and companies. Figuring out how to adapt these to solve our customers' and our nation's problems is extremely challenging and rewarding. I am also spending more time engaging with students and our youth, encouraging them to strengthen their STEM skills, and to think about careers in STEM. This has been equally exciting. It's fascinating how much smarter and aware they are than I was at their respective ages.

**My life is:** My life is a blend of service to, and support from my family, my industry, my community, my employees and customers, and our nation. I am truly a product of what is good about our country, and though there is much more we can all do to be better and to treat each other better, I am very optimistic about the future.

---

## *References*


1. George, B. *True North: Discover Your Authentic Leadership.* Available at: http://www.servicesmag.org/industry/business/item/397-true-north-discover-your-authentic-leadership
2. Pew Research Center. *Beyond Distrust: How Americans View Their Government.* 2015.
3. Mori, I. *Politicians Trusted Less Than Estate Agents, Bankers and Journalists.* Available at: http://www.people-press.org/2015/11/23/1-trust-in-government-1958-2015/, accessed on October 13, 2016.
4. Goleman, D., et al. *The New Leaders—Transforming the Art of Leadership into the Science of Results.* London, UK: Time-Warner, 2002.
5. Bennis, W. The secrets of great groups. *Leader to Leader,* 1997: 29–33, 1997.

6. Bush, P.M. *Transforming Your STEM Career through Leadership and Innovation: Inspiration and Strategies for Women*. Academic Press, 2012.
7. Advise America. Visionary Leadership Examples. Available at: http://www.adviseamerica.com/visionary-leadership-examples/
8. Wikipedia. Mahatma Gandhi. Available at: https://en.wikipedia.org/wiki/Mahatma_Gandhi
9. Myatt, M. *10 Communication Secrets of Great Leaders*. Available at: http://www.forbes.com/sites/mikemyatt/2012/04/04/10-communication-secrets-of-great-leaders/
10. Hall, A. *The Top Communication Traits of Great Leaders*. 2012. Available at: http://www.forbes.com/sites/alanhall/2012/07/06/the-top-communication-traits-of-great-leaders/#3e215b0b7879
11. Cohen, S.L. Effective global leadership requires a global mind-set. *Industrial and Commercial Training*, 42(1): 2020: 5–10, 2010.
12. Goldsmith, M.I., Fisher, S., Waterman, R., and Johnson, S.L. Salutatory control of isometric growth in the zebrafish caudal fin is disrupted in long fin and rapunzel mutants. *Developmental Biology*, 259(2): 303–317, 2003.
13. Wikipedia. *Thinking outside the box*. Available at: https://en.wikipedia.org/wiki/Thinking_outside_the_box
14. Quora. *How Do I Create the Ability to Think Out of the Box?* Available at: https://www.quora.com/How-do-I-create-the-ability-to-think-out-of-the-box
15. Wikipedia. Galileo Galilei. Available at: https://en.wikipedia.org/wiki/Galileo_Galilei
16. Maxwell, J.C. *The 21 Irrefutable Laws of Leadership*. Thomas Nelson Publishers, Nashville, 1998.
17. Goleman, D. What makes a leader. *Harvard Business Review*, November–December, 93–102, 1998.
18. Goldsmith, A., et al. *Global Leadership: The Next Generation*. Upper Saddle River, NJ: Prentice-Hall Financial Times, 2003.
19. Gumbel, P. Big Mac's local flavor. *Fortune*, 115–118, 2008. Available at: http://archive.fortune.com/2008/04/29/news/companies/big_macs_local.fortune/index.htm, accessed on October 13, 2016.
20. Reiche, S., et al. *What Is Global Leadership?* Business & Innovation. March 14, 2013. Available at: http://www.worldfinancialreview.com/?p=859
21. Bellet, P.S. and Maloney, M.J. The importance of empathy as an interviewing skill in medicine. *JAMA*, 226(13): 1831–1832, 1991.
22. Batson, C.D. These things called empathy: Eight related but distinct phenomena. In J. Decety and W. Ickes. (Eds.), *The Social Neuroscience of Empathy*. Cambridge, MA: MIT Press, 2009, pp. 3–15.
23. Hatfield, E., Cacioppo, J.L., and Rapson, R.L. Emotional contagion. *Current Directions in Psychological Sciences*, 2(3): 96–99, 1993.
24. Wikipedia. Social skills. Available at: https://en.wikipedia.org/wiki/Social_skills
25. Psychology Today. *How to Increase Your Emotional Intelligence—6 Essentials Six Ways to Increase Your Emotional Intelligence*. 2014. Available at: https://www.psychologytoday.com/blog/communication-success/201410/how-increase-your-emotional-intelligence-6-essentials, accessed on October 13, 2016.
26. Goleman, D. Leadership that gets results. *Harvard Business Review*, March–April, 2000, pp. 78–90.

27. Wikipedia. We choose to go to the Moon. Available at: https://en.wikipedia.org/wiki/We_choose_to_go_to_the_Moon
28. Money-Zine. *Coaching Leadership*. 2015. Available at: http://www.money-zine.com/career-development/leadership-skill/coaching-leadership/
29. Lewin, K., et al. Patterns of aggressive behavior in experimentally created social climates. *Journal of Social Psychology*, 10, 271–301, 1939.
30. Wikipedia. Dwight D. Eisenhower. Available at: https://en.wikipedia.org/wiki/Dwight_D._Eisenhower
31. Assessing your leadership style. Available at: http://kkytbs.org/searchresults.html?q=Assessing%20Your%20Leadership%20Style, accessed on October 13, 2016
32. Dyer, J.H., et al. The Innovator's DNA. *Harvard Business Review*. 2009. Available at: https://hbr.org/2009/12/the-innovators-dna, accessed on October 13, 2016.
33. Krishnan, V. and Ulrich, K.T. Product development decisions: A review of the literature. *Management Science*, 47(1), 1–21, 2001.
34. Kahn, K.B. *The PDMA Handbook of New Product Development*. 3rd ed. Hoboken, NJ: Wiley, 2013, p. 21.
35. Giorgio, T. *Bill Gates and Warren Buffett at Columbia Business School*. 2016. Available at: http://giorgiotomassetti.blogspot.com/2009/11/bill-gates-and-warren-buffett-at.html, accessed on November 17, 2016.
36. Schuster, E. Dimension. *Leadership during Times of Crisis*. 2015. Available at: http://www.astc.org/astc-dimensions/leadership-during-times-of-crisis/.

# *chapter five*

# *Establishing the vision as an engineering leader*

## *The foundation of vision*

A popular proverb states that "Without a vision the people perish" [1], and this also can be applied to an organization, a project, or a team. A vision is a powerful tool for motivating, inspiring, and moving toward a lofty yet attainable reality, something more than a dream—a goal to work toward. Vision is the key driver of leaders, as it takes their steadfast belief and dedication to see it come to fruition. A vision, no matter what it is, propels a leader forward and into action, and thereby compels him/her to inspire, empower, and encourage others. No matter how powerful the vision is, it must have at its "root" passion. Passion becomes the fuel for career success, personal achievement, and fulfilling one's purpose. Passion extends beyond career success perspectives, and when linked to our core values and ideals, it serves as a continuous source of energy that offers continuous motivation, resilience, and hope.

The vision essentially has a foundation with which it is linked to our passion. In previous generations, many technical leaders were reluctant to share or express passion, but fortunately this is changing. Passion is important not only for inspiration in the individual leader but also to inspire and empower those who are leading. Thus, passion has a way of creating a sense of trust and deeper relationships between a leader and the team [2]. This is even more evident in dangerous or high-risk environments. Given the significance of many engineering project outcomes, it is very important that all engineers understand the importance of passion and pursue an understanding of it in our personal and professional lives as this will directly impact the individual, team, and organizational outcomes.

## *What matters to you?*

We become the best leaders by doing things we are passionate about. Take the time to study your interests, the things that make you angry, what makes you happy, and the things you regularly find yourself focusing on at work and at home. Doing this will lead to an understanding of what *really* matters to you. What you decide needs to apply to your own personal beliefs and interests, and only then this can be aligned with professional ambitions. This may be a matter of technical problems you want to solve, a team of people you are leading, or even a technology that you want to create to save lives in developing nations. This is the key part of establishing the process of creating a vision.

## Developing the visionary leader in you

The Great Recession of 2008 highlighted the absence and criticality of global technical leaders. Engineers and technical professionals are needed to initiate, innovate, and lead change to meet the current and emerging needs of the global society. As such, it is imperative that individuals understand *how* to develop as leaders. These principles were initially shared in 2012 in an engineering leadership text focused on women in STEM [3]. However, these characteristics apply to all aspiring professionals.

### Establishing your vision and mission statement

I quite clearly remember being a young, single mother with ambition of a thriving profession and good life for my little daughter. Even though I was making use public assistance (welfare), struggling with my academics, and was unsure of what my future had in store, I endeavored to cling unto a vision of a better life. Each week I thought about how my life would be when I won't need to depend on welfare and how I would go on vacations, become a successful engineer, and build a great home for my daughter. These visions were a supply of motivation and support to me during the lowest peak of my life and they were the basis that stirred me onward to the realization of my ambitions. At present, I still do the same thing in taking into consideration the next stage of my career. My leadership model has been polished, and though my motivation has altered, the course of action for getting there remains the similar: produce a vision, a mission, and implement the plan!

When you are developing a team vision statement, you are always dealing with the future. You are describing the destination towards which you are aiming. You are not setting out how you are going to arrive—that comes much later. Your vision statement is never static, but always dynamic and is designed to set free an organization's energy. Here are examples of Vision Statements:

**Disneyland:** *Create a place for people to find happiness and knowledge*
**Ford:** *Produce a car that everyone can afford*
**Girl Scouts:** *Help a girl reach her highest potential*
**Cirque du Soleiol:** *Invoke the Imagination; Provoke the Senses; Evoke the Emotions*
**Zappos:** *The online service leader* [4]

Note the breadth of these vision statements. They are brief but dynamic, succinct statements for what the organization wants to achieve and be known for in the industry.

Regardless of whether you come to a decision to subscribe to leadership theories of authentic, transformational, or any other theory, an easy method to commence on your journey into leadership is to set up a vision and mission statement for your own personal career goal. This breakdown helps to give you a better perception of your areas of strengths, what your interests really are, as well as the opportunities you are faced with, in addition to the threats which you must surmount to accomplish a specific goal. Make an effort to be as goal oriented as possible about yourself. Not simply should you just recognize your strengths and weaknesses within this analysis, but you should in addition list out your assets (not just your financials!) and your liabilities. If you have a number of close confidants whose judgment you rate high and whom you trust, show them your list and request for their observations. Bear in mind that it is not always the big stuff that enlighten you about yourself. You may perhaps have raised a lot of money for a cause, which may well surmise that you are a good motivator; nevertheless, if your workspace at home is constantly in disarray, you may not be an excellent organizer. Except you are a saint, your list of weak points will be equivalent to your strengths.

The growth of a vision statement takes time; nonetheless, the investment of time will be well worth the exertion in the long term. Following the development and growth of your vision, you will expectedly see your mind-set improved and the occurrence of your dissatisfaction with little matters in life begin to reduce. That is for the reason that your mind will be on "higher" things, particularly your life aspirations. A vision is in relation to a higher calling and a vision-driven living is a stimulating and impactful way to live.

Your vision statement should also comprise the essential areas of your life. It is a written depiction of your desired future life as you perceive it in your mind. There are no standards about the "appropriate" layout or length of a vision statement. However, the more comprehensive and exact your vision is, the more it will be attached to you and this will increase the likelihood of establishing a clear visualization for the attainment of the vision. Some helpful questions to ask yourself while developing your personal vision statement include the following:

- What do I believe my purpose is as a leader?
- What are my most important values?
- What are the things that I really enjoy?
- What brings me happiness/joy?
- What are the issues and causes that I care deeply about?

- What are my primary strengths?
- What are the things that I'd like to stop doing or do as little as possible?
- How can my purpose best serve the people that matter most to me?

Table 5.1 can be used to categorize your answers to these questions.

After you have completed these questions, consider the holistic perspective that results from the things that bring you happiness, your values, strengths, and desire to serve. Capture this perspective by summarizing your thoughts in Table 5.2 [3].

## *Your personal mission statement*

After the completion of your vision statement, the next step of action is the development of a mission statement. In quintessence, the mission statement "operationalizes" your vision and provides a brief platform to support your thoughts and activities from your personal vision statement. Therefore, creating the mission necessitates the assessment of the vision as it concerns activities, resources, and the effect on the external groups. In other words, your mission statement is the way through which you will manifest your personal vision in your everyday life. It may be a small number of words or quite a few pages, but it is not just a "to do" list. It depicts your exceptionality and must speak to you strongly about the person you are, the person you are becoming, and the necessary actions to move you forward.

***Table 5.1*** Personal vision statement tool #1

| | | | |
|---|---|---|---|
| Things I really enjoy doing | Issues or causes I care deeply about | The two best moments of my past week | Three things I'd do if money or time was not an issue |
| What I'd like to stop doing or do as little as possible | My most important values (Circle) | Things I can do at the good-to-excellent level | Things I can do to serve the people that matter the most to me |

***Table 5.2*** Personal vision statement tool #2

1. Based on my personal research, these are the main things that motivate me/bring me joy and satisfaction:
2. My greatest strengths/abilities/traits/things I do best:
3. At least two things I can start doing/do more often that use my strengths and bring me joy:
4. This is my personal vision statement for myself (in 50 words or less):

Based on the development of your personal vision, you can at present commence the improvement of your mission statement. While preparing the mission statement, it is useful for the decision makers to reflect on the following subjects:

- What has to be accomplished to translate vision into action?
- How the activities will be carried out and what are the necessary resources?
- The beneficiaries of the mission being attained.

## *What makes a good vision and mission statement?*

- A good vision and mission statement is concise and inspirational.
- It is easy to memorize and repeat.
- It should be clear, engaging, and realistic, and should describe a bright future.
- It should furthermore state your intentions, summarize your values, and demonstrate your commitment to living up to these values.

**Example:**

> *My mission is to help project managers transform into impactful project leaders.* [5]

Once your mission statement is completed, the very next step is to create a plan that can be put into action so as to begin moving in the direction of the realization of the vision. This sequence of actionable items will develop into your personal road map.

Finalizing your vision and mission statements will probably take various iterations; therefore, it is prudent to take a few days to cultivate and think over the vision as it concerns your general life plan. To have a personal vision does not connote that your life will be transformed immediately. Nevertheless, with a personal vision statement, mission statement, and dedication, your life will surely be transformed. Your personal mission statement presents the road map and actions to get you there [3].

## *Creating your leadership development plan*

The growth of a string of developmental actions to improve your competence should be based on your personal vision, professional aspirations, private life, and access to resource. Your leadership development plans should comprise an array of techniques together with actions within your place of employment, professional society, or community-based activities and individual actions. As you produce your developmental activities, appraise each utilizing the SMARTER model where every activity needs to meet the following criteria [6].

The SMARTER system can be useful for planning your career goals:

- **Specific:** Be as clear as you can and avoid ambiguous statements.
- **Measurable:** So you can see what you have achieved.
- **Achievable:** Provides motivation, but also keep your goals reachable.
- **Realistic:** Be ambitious for sure, but avoid the realms of fantasy.
- **Timely:** Create time frames for completing steps, for example, doing short courses or talking with someone about the skills required for a particular job.
- **Empowering:** Make sure your goals feel right for you and help you make the changes you want.
- **Reviewable:** Do not set your goals in concrete; be flexible.

Write clearly defined, short statements that you can work toward. If you are unable to identify a specific job you want, indicate your goals in more general terms. This is all part of good career strategy foundations. But remember: the more specific you will be, the easier it will be for you to plan [7].

To commence the development of your plan, reflect on the ideas given in Figure 5.1a through c [8]. This string of templates can be utilized as a brief reference to control your vision, goals, actions, and plan to track results. The templates should be utilized as follows:

- Figure 5.1a—Vision, Mission, and Goals: Insert your vision statement, mission statement, and goals that you have identified in the previous sections.
- Figure 5.1b—Strength Enhancement: Here, you will list the areas that you want to develop within your strengths. If you are doing this as an activity apart from your job, consider asking a mentor or a trusted friend to serve in the manager's role. Alternatively, it would be a demonstration of leadership on your part to present such a plan to your manager, even if the organization presently doesn't have an employee development program. Be careful in these circumstances, as your manager may or may not be friendly to your forward thinking approach with respect to your profession.
- Figure 5.1c—Developmental Needs: Here, you will list the top two areas that you want to focus on developing in the next six months. Your individual development plan should be updated at least every six months and the status of development should be documented within each area.

**Individual development plan**

Name: ______________________ Manager: ______________________
Position: ______________________ Date: ______________________
Date in current position: ______________________

| **Section A: Career plan** |
| --- |
| Personal mission statement |
| |

| **Short-term career goals (1–2 years)** | |
| --- | --- |
| Area of interest/position title | Competencies/skills/knowledge needed: (areas I need to develop) |
| | |
| | |
| | |

| **Long-term career goals (3–2 years)** | |
| --- | --- |
| Area of interest/position title | Competencies/skills/knowledge needed: (areas I need to develop) |
| | |
| | |
| | |

It represents a template that can be used to created an individual development plan for professional growth.

(a)

***Figure 5.1*** Individual development plan. (From PHS Mentoring Program for Engineers. *Individual action plan/individual development plan (IAP/IDP)*. Available at: https://sites.google.com/site/phsmentoringforengineers/home/individual-development-plan, accessed on November 18, 2016. With permission.) *(Continued)*

**Individual development plan**

| **Strength to leverage:** select at least one strength to continue to build upon | | | **Area of focus:** | |
|---|---|---|---|---|
| Critical Behavior/ Goals What specific behaviors do I need to model or exhibit in this competency or skill? | Developmental Activities/Action Steps (assignment, coaching, formal training) Remember SMART | Manager's Role (or involvement of others if applicable) | Target Dates/Mile-stones | Result/Outcomes How have I succeeded in adapting my behavior or learning new skills? (Provide examples) |
| | | | | |

(b)

| **Area to develop:** focus on areas to develop that are critical to your performance; select 1 or 2 areas to work on at one time | | | **Area of focus:** | |
|---|---|---|---|---|
| Critical Behavior/ Goals What specific behaviors do I need to model or exhibit in this competency or skill? | Developmental Activities/Action Steps (assignments, coaching, formal training) Remember SMART | Manager's Role (or involvement of others if applicable) | Target Dates/Mile-stones | Results/Outcomes How have I succeeded in adapting my behavior or learning new skills? (Provide examples) |
| | | | | |

(c)

***Figure 5.1 (Continued)*** Individual development plan. (From PHS Mentoring Program for Engineers. *Individual action plan/individual development plan (IAP/IDP)*. Available at: https://sites.google.com/site/phsmentoringforengineers/home/individual-development-plan, accessed on November 18, 2016. With permission.)

## Attributes of the global engineering leader

The significance of understanding an engineer's roles as a leader and applying that vision, mission, and road map to achieve the engineering outcomes is the goal of engineering leadership. The implications are growing due to the connectivity and impact of engineering outcomes. In a workshop at the Massachusetts Institute of Technology (MIT), the *Capabilities of an Engineering Leader* emerged from the consensus of a group of stakeholders [9]. The stakeholders group was consisted of a diverse collection of engineering leadership representatives including alumni, students, faculty, leaders from industry, military leaders, community leaders, and those from other leadership programs at MIT. Given the significant shift in the connectedness and global reach of the engineering profession, an additional quality of "Servant Leadership" is included with the five categories offered by the MIT workshop. The resulting categories for success as an Engineering Servant Leader are as follows.

### The attitudes of leadership—core personal values and character

Students who aspire to become leaders must contemplate on their viewpoint and manner, and also should develop a sense of dependability and personal ability that outline groundwork for effective leadership. For effective engineering leaders, these include:

- Initiative
- Decision-Making in the Face of Uncertainty
- Responsibility, Urgency, and Will to Deliver
- Resourcefulness, Flexibility, and Change
- Ethical Action, Integrity, and Courage
- Trust and Loyalty
- Equity and Diversity
- Vision and Intention in Life
- Self-Awareness and Self-Improvement

### Relating

A leader should grow key interactions and network with people within and across organizations. These comprise giving attention to others to comprehend their views and promotion for your position. For efficient engineering leaders, these specialize to:

- Inquiring and Dialoguing
- Negotiation, Compromise, and Conflict Resolution

- Advocacy
- Diverse Connections and Grouping
- Interpersonal Skills
- Structured Communications

### *Making sense of context*

This includes the ability of a student to make sense of the world around him/her, understanding the framework in which the leader is working—creating an intellectual map of the multifaceted environment, and explaining it simply to others. For effective engineering leaders, these specialize to:

- Awareness of the Societal and Natural Context
- Awareness of the Needs of the Customer or Beneficiary
- Enterprise Awareness
- Appreciating New Technology
- Systems Thinking

### *Visioning*

This includes producing resolute, persuasive, and transformational descriptions of the future and recognizing what may and ought to be. For effective engineering leaders, these specialize to:

- Identifying the Issue, Problem, or Paradox
- Thinking Creatively, and Imagining and Communicating Possibilities
- Defining the Solution
- Creating the Solution Concept

### *Delivering on the vision*

This includes pioneering innovation through planning, procedures, and methods in an attempt to deliver on the vision and to advance from concept to innovation, creation, and execution, that is, to get the engineering completed. For effective engineering leaders, these specialize to:

- Building and Leading an Organization and Extended Organization
- Planning and Managing a Project to Completion
- Exercising Project/Solution Judgment and Critical Reasoning
- Innovation
- Invention
- Implementation and Operation

## Technical knowledge and reasoning

A profound operational understanding of the skill or discipline is indispensable to the successful implementation of an engineering leadership. Although usually enhanced in the average curricular course of study, this knowledge is no less important for an engineering leader. It comprises the capability to comprehend, decompose, and recombine diverse fundamentals of a technological problem through the use of a profound understanding of technological knowledge.

## Servant leader attitude

"Servant Leadership" is an approach to leadership, with strong philanthropic and moral undertone that requests and calls for leaders to be caring about the needs of their group and identify with them; they should take care of them by ensuring they become healthier, wiser, freer, and more self-sufficient so that they too can become servant leaders [10]. The engineering leader impacts the lives of others through products, technology, and services and ultimately "serves" the customer with these products. It is prudent for the engineering community to consider the value of servant leadership attributes for those we lead and those that our products, services, and talents eventually reach.

## Ten characteristics of servant leadership

Following many years of careful consideration of Greenleaf's original writings, the following set of characteristics central to the development of servant leaders was extracted [11] from the comprehensive discussion of Servant Leadership. Each of these qualities as described by the author is listed below:

1. **Listening**
   Leaders have usually been appreciated for their communication and decision-making skills. While these are also important skills for the servant-leader, they need to be reinforced by a deep commitment to listening intently to others. The servant-leader seeks to identify the will of a group and helps clarify that will. He or she seeks to listen receptively to what is being said. Listening, coupled with regular periods of reflection, is essential to the growth of the servant-leader.

2. **Empathy**
   The servant-leader makes every effort to comprehend and empathize with others. People need to be acknowledged and known for their exceptional and distinctive state of mind. One presumes the good objective of colleague and does not snub them as people, even if one discovers it essential to decline accepting their actions or performance.

3. **Healing**
   One of the immense strong points of servant-leadership is the possibility for healing one's self and others. Several people have broken state of mind and have undergone a variety of emotional hurts. Even though this is part of being human, servant-leaders recognize that they also have a chance to "help make whole" those with whom they come in contact.
4. **Awareness**
   General awareness, and in particular self-awareness, strengthens the servant-leader. Awareness also aids one in understanding issues involving ethics and values. It lends itself to being able to view most circumstances from a more integrated, holistic position.
5. **Persuasion**
   A further feature of servant-leaders is a major dependence on persuasion rather than positional authority in making decisions within an organization. The servant-leader seeks to influence others rather than force obedience. The servant-leader is successful at building consensus within groups.
6. **Conceptualization**
   Servant-leaders seek to foster their abilities to "dream great dreams." The capacity to look at a problem (or an organization) from a conceptualizing perspective means that one must think beyond day-to-day realities. For several managers this is a feature that requires discipline and practice. Servant-leaders are called to look for a delicate balance between conceptual thinking and a day-to-day focused approach.
7. **Foresight**
   Foresight is an attribute that enable[s] the servant-leader to comprehend the lessons from the past, the realities of the present, and the likely consequence of a decision for the future. It is also deeply rooted within the intuitive mind. Foresight remains a largely unexplored area in leadership studies, but one most deserving of careful attention.
8. **Stewardship**
   Peter Block has described stewardship as "holding something in trust for another." Servant-leadership, like stewardship, assumes first and foremost an obligation to serving the needs of others. It also emphasizes the use of openness and persuasion rather than control.
9. **Commitment to the Growth of People**
   Servant-leaders consider that people have an inherent value further than their tangible contributions as workers. As a consequence, the servant-leader is deeply committed to the development of each and every individual within the organization. The servant-leader

recognizes the marvelous responsibility to do everything feasible to cultivate the growth of employees.

10. **Building Community**
The servant-leader senses that much has been lost in modern human history as an end result of the change from local communities to large institutions as the major shaper of human lives. This consciousness causes the servant-leader to seek to recognize some way for building community among those who work within a given organization. Servant-leadership advocates that true community can be produced among those who work in businesses and other institutions.

These 10 characteristics of servant leadership provide a detailed overview of what the application of servant leadership should look like in practice. Many of these attributes are found in other leadership approaches and should also be integrated into effective engineering leadership.

## *Vision, leadership, and entrepreneurship*

It is an exciting time to be an engineer as the opportunities for transitioning one's technical skills and passion into lucrative entrepreneurial ventures have never been greater in the global society. The National Academy of Engineering recognizes this, as a recent journal stated the following [11]:

> It is no longer enough to come out of school with a purely technical education; engineers need to be entrepreneurial in order to understand and contribute in the context of market and business pressures. For engineers who start companies soon after graduation, entrepreneurship education gives them solid experience in product design and development, prototyping, technology trends, and market analysis [11]. These skills are just as relevant for success in established enterprises as they are in startups; students with entrepreneurial training who join established firms are better prepared to become effective team members and managers and can better support their employers as innovators. Entrepreneurship education teaches engineering students in all disciplines the knowledge, tools, and attitudes that are required to identify opportunities and bring them to life. [10]

This is exciting indeed, but how exactly does an engineering student or practicing engineer acquire the necessary entrepreneurial skills to

*Table 5.3* Respondents' mean and standard deviation on important personal characteristics for entrepreneurs to become successful, N = 122

| N | Personal Characteristics | Mean | SD | Result |
|---|---|---|---|---|
| 1 | Optimism | 4.46 | 0.63 | High |
| 2 | Originality | 4.66 | 0.55 | High |
| 3 | Vision | 4.76 | 0.50 | High |
| 4 | Discipline | 4.12 | 0.98 | High |
| 5 | Endurance | 4.50 | 0.67 | High |
| 6 | Initiative | 4.64 | 0.63 | High |
| 7 | Self-confidence | 4.47 | 0.61 | High |
| 8 | Motivation | 4.45 | 0.66 | High |
| 9 | Courageous | 4.66 | 0.61 | High |
| 10 | Flexibility | 4.53 | 0.65 | High |
| 11 | Willingness to learn | 4.47 | 0.61 | High |
| 12 | Creativity | 4.52 | 0.81 | High |
| | **Grand Mean** | **4.52** | **0.66** | **High** |

*Source:* From Ezenwafor, J.I., and Okoli, C.I., *J. Emerg. Trends Econ. Manag. Sci.*, 5(7), 51, 2014. With permission.

launch an enterprise or product? First, it is important to *know* what are the necessary qualities to be a successful technical entrepreneur. In a study by Ezenwafor and Okoli [11], 149 managers were surveyed to determine the most important entrepreneurial characteristics. As expected, "vision" or "being a visionary" was found with the highest rating for any of the characteristics in the analysis. The most important personal characteristics for success as an entrepreneur are given in Table 5.3.

Thus, it is prudent for the hopeful engineering entrepreneur to focus on the development of these qualities. The list seems to be exhaustive and could appear daunting to the young engineer; however, these skills can be acquired and grown at different times and phases of one's career. There are a number of ways to acquire, grow, and enhance these characteristics, including formal education and training, experiential learning, entrepreneurial development programs, and personally initiated learning activities (i.e., reading books, creating opportunities in personal experiences to develop these qualities).

## Conclusion

Successful leadership includes the capacity to create an inspiring mission and vision statements, create trust within an organization, and get out the best in people. Vision is the core of leadership and key to strategy. A leader's responsibility is to create the vision for the organization in such a way that it will fit into place both the imagination and the energies of its people.

An effectual leader discerns that the decisive task of leadership is to create human energies and human vision. The vision must be tied to the firm's morals, and the leader must make this link in such a way that the organization can comprehend, grasp, and support. Vision moves the organization; values stabilize the organization. Vision looks to the future, whereas values look to the past.

If your vision statement opens people's eyes to what is achievable, and motivate them to work toward accomplishing that goal, then it has served its functions well. Of course, it is the leader who must continuously and vigorously remind and unite people to the vision.

---

**Case study: Elon Musk**

Elon Musk is perhaps the most eminent visionary of our time. The engineer, inventor, and entrepreneur is the cofounder of PayPal, founder and CEO of SpaceX, CEO and product architect of Tesla Motors, chairman of Solar City, and cochairman of Open AI, among other projects.

While his resume is varied, the thing that ties all of Musk's endeavors together is his vision. This vision—to aid mankind by promoting sustainable energy consumption, reducing greenhouse gas emissions, and if necessary, colonizing Mars—is being carried out through several projects in which Musk has invested much of his time and resources.

Musk's ideas have been called outlandish at times. An example of this is his idea to build a 700-mile-per-hour Hyperloop with Star Trek-style transportation pods designed to carry passengers from Los Angeles to San Francisco in less than 30 min on a "cushion of air." Musk released a 57-page document in 2013 detailing his idea, where he admitted it wouldn't be easy: "Short of figuring out real teleportation, which would of course be awesome (someone please do this), the only option for super-fast travel is to build a tube over or under the ground that contains a special environment. This is where things get tricky" [12].

Although his vision for the fastest transportation system of our time was considered impossible by many, in early 2016, thousands of engineering students gathered to showcase their designs for the Hyperloop Pod Competition and to vie for a chance to be a part of this exciting enterprise.

Musk doesn't just want to improve life on earth through faster transportation. He is prepared to transport humans to Mars in the future should the Earth no longer be inhabitable.

In an interview with *National Geographic,* Musk's biographer, Ashlee Vance, echoed popular opinion about Musk's company, SpaceX, saying, "It's nuts! It should not exist at all! Usually it's governments that try to do these things" [13]. Many were skeptical that SpaceX would be able to compete with the likes of Lockheed and Boeing, but Musk hired as many smart, talented engineers as he could find and set out to build a rocket. Today SpaceX launches satellites for countries and private companies and resupplies the international space station, while preparing for the day when a manned mission to Mars is imminent.

When asked in an interview with the BBC's technology correspondent about what unites his three primary interests—SpaceX, Tesla, and Solar City—Musk responded, "What I'm trying to do is to minimize future existential threats or take whatever action I can to ensure the future is good. I didn't expect these companies to succeed. I thought they would most likely fail [...] I think there are some things that are important for the future, sustainable energy ... sustainable transport ... ultimately becoming a multi-planet species ... and those are the things that make me like the future and feel inspired about the future, whereas if those things don't happen, the future I think looks quite dim" [14].

Elon Musk sees a world where cars drive automatically, public transportation is swift and efficient, and the world runs on alternative and renewable energy resources. He sees global warming endangering an ever-growing population on a planet that is going to be uninhabi one day. He worries about the rise of technology and artificial intelligence, and invests in artificial intelligence companies in order to keep an eye on them. His concern for the survival of the human race has been termed *obsessive* by his colleagues. This can be considered the mark of a true visionary, and the persistence and drive to make his visions a reality are all part of what makes Elon Musk a leader in his field.

## QUESTIONS

1. What engineering leadership characteristics were demonstrated in this case study?
2. How were these characteristics applied to deliver the intended impact?
3. What was the technical, societal, or environmental impact due to the application of the leadership principle(s) identified above?

---

**Real profiles in engineering leadership**

**Name:** Duy-Loan Le (phonetically ZLON LEE)

**Current position or field of expertise:**

- Texas Instruments (TI) Senior Fellow (retired), Semi-conductor
- National Instruments Inc., Board Director, Chair of Compensation Committee
- eSilicon Corp., Board Director, Chair of Technology Committee
- Medigram Inc., Board Director

**What I like most about my position:** For 33 years at Texas Instruments (TI), I had the opportunity to travel the world, train engineers in multiple countries, work with our global work force, collaborate with our suppliers and partners in multiple continents, develop leading edge technologies, solve very complex problems, design and manufacture useful products to benefit society, mentor people and help them to succeed along the way, and learn from so many human beings I have met from all walks of life. In parallel, my board positions at three other companies allow me to grow intellectually, contribute at the highest corporate leadership level, and interact with executives beyond TI environment. In other words, I really enjoy having not just one company to grow with but four companies to contribute to and learn from. I particularly appreciate having the influence to create synergy between what I do for the companies and what I want to give back through my philanthropic activities in the field of education and STEM. I really like this win-win strategy and treasure the autonomy that comes with the sphere of influence.

**Who or what has made a difference in my career:** When I was in high school, I dreamed of becoming a doctor: I think there is nothing more beautiful than making a difference by saving lives with your knowledge and skill. However, it was not a practical path given the fact that I started life in America countryless, homeless, fatherless, penniless, and speechless (a play on word as I spoke no English). The reality staring me in the face at 13 was: I need to make money to help my family! I hurried through high school and graduated valedictorian at 16, pursued engineering and accelerated my education to graduate Magna Cum Laude at 19, and started my career at TI as a DRAM design engineer. DRAM is Dynamic Random Access Memory.

Once an engineer, what drove me to excellence is the desire to honor the country that gave me birth, uphold the trust of Mr. Lionel White who hired me into TI, carry out the dream of making a difference with my STEM knowledge (vs medical), and live up to my own standard. That standard is deeply entrenched in the work ethics of my upbringing in Vietnam working side by side with my father and rooted with a deep sense of gratitude for the land I love so much called America and her generous people.

**I felt like quitting when:** In a career that lasted 33 years at Texas Instruments, there were moments that I failed, challenged to the limit, and wondered "why is life so hard?" Adding to the complexity of the job at TI, I was determined to raise my boys not only as American but also as Vietnamese, commuted between Dallas and Houston half of my career, cofounded two nonprofits and served on the founding boards of two others (all four are still in operation today), elected to two corporate boards and two university boards, and give keynotes by invitation. Keeping all the balls in the air can be nerve-racking to say the least … but it was all my choosing! Physically it is painful for the body, intellectually I am always challenged, but emotionally I am very happy (almost always, that is). I am grateful to Texas Instruments for making it possible to do all of the above to begin with … you know not every company is that trusting.

I never felt like quitting because I have the understanding and patience from my husband (Tuan), selfless support from my older sister (Amina), good character and nature of my two sons (Dan and Don), generous help from many close friends, and competency of all assistants in various companies. The fact remains: there is no superwoman … It does take a village!

**My strategies for success are:** I have principles that guide me through life and it has been a good journey.

- *Be authentic*: define what makes you happy and do not live the life others have or want for you.
- *Execute to excellence*: big or small job, important or mundane, for money or for pro-bono. Do it with passion because of self-respect.

- *Have thick skin*: do listen to others' inputs but trust your instinct. Do not waste time feeling guilty. Channel energy into productivity.
- *Help people*: make it easy for others to ask for help. Do it not in exchange for good karma but because it is the right thing.
- *Take risks*: adventure and make mistakes as long as it is not the same one. Have the courage to forgive yourself but *never* lose the lesson.

**I am excited to be working on:** I value autonomy over money. I retired at 52 to gain more time for myself. I want to be able to wake up when I want, spend time on things that I enjoy, walk not run, continue to make a difference with my means, and answer to no one! When I was at Texas Instruments, I treasured most the autonomy that TI had entrusted in me and that very trust made it impossible not to be 105 percent devoted to the company. I am really excited to have my autonomy without the moral obligation to work ninety hours per week like the last four decades.

**My life is:** I am truly grateful to this country for the freedom and opportunities to do what I want, be loved by so many, and highly respected in the global community. I cherish the three lives in parallel:

- *The personal life*: with my husband of 33 years and my two precious sons who are so independent and yet so affectionate. They are great human beings and this makes me very appreciative and proud.
- *The professional life*: with Texas Instruments and all the boards I contribute to … intellectual stimulation is very important to me. It keeps me young.
- *The philanthropic life*: to give back and make a difference with my time, money, skills and connections worldwide. Giving back is about playing fair because I have received so much.

To be able to live three lives in one lifetime is amazingly fortunate and I must continue to live up.

---

## References

1. *The Holy Bible*, Proverbs 29:18, the King James Version translation.
2. Eisenberger, N., & Kohlrieser, G. Lead with your heart, not just your head. *Harvard Business Review*, November 16, 2012.

3. McCauley-Bush, P. (2012). *Transforming Your STEM Career through Leadership and Innovation: Inspiration and Strategies for Women*. Academic Press: Elsevier, London, UK.
4. Make A Dent Leadership. *Developing a Team Vision Statement: The Elements of a Good Vision Statement*. Available at: http://www.makeadentleadership.com/developing-a-team-vision-statement.html, accessed on October 13, 2016.
5. Madsen, S. (2013). *How to Create a Personal Mission and Vision Statement for the Year.* Available at: https://www.liquidplanner.com/blog/create-personal-mission-vision-statement-year/, accessed on October 13, 2016.
6. State Government of Victoria. *A Career Plan*. Available at: http://careers.vic.gov.au/exploration/a-fair-workplace, accessed on October 13, 2016.
7. Gordon Leadership Institute. Workshop Preliminary Report. 2012. http://news.mit.edu/2012/educational-exchange-program-0110
8. PHS Mentoring Program for Engineers. *Individual action plan/individual development plan (IAP/IDP).* Available at: https://sites.google.com/site/phsmentoringforengineers/home/individual-development-plan, accessed on November 18, 2016.
9. Northouse, P. G. (2004). *Leadership: Theory and Practice*. Thousand Oaks, CA: Sage, pp. 308–309.
10. Byers, T., Seelig, T., Sheppard, S., & Weilerstein, P. (2013). Entrepreneurship: Its Role in Engineering Education. *The Bridge*, 43(2), 35–40.
11. Ezenwafor, J. I., & Okoli, C. I. (2014). Assessment of Personal Characteristics Needed to Become Successful Entrepreneurs by Managers of Small and Medium Enterprises in Anambra and Enugu State, Nigeria. *Journal of Emerging Trends in Economics and Management Sciences*, 5(7), 51.
12. FoxNews TECH Online. *Crowdfunding called to power Hyperloop high-speed transport*. Available at: http://www.foxnews.com/tech/2013/08/29/crowdfunding-called-to-power-hyperloop-high-speed-transport.html, accessed on November 11, 2016.
13. Worrell, S. *Elon Musk, a man of impossible dreams, wants to colonize Mars*. Available at: http://news.nationalgeographic.com/2015/06/150628-tesla-paypal-elon-musk-technology-steve-jobs-silicon-valley-electric-car-ngbooktalk/, accessed on November 18, 2016.
14. Gray, K. (2015). An Astronaut's Guide to Life on Earth, by Chris Hadfield Elon Musk: Tesla, SpaceX, and the Quest for a Fantastic Future, by Ashlee Vance: New York, NY: Back Bay Books, 320 pp.

# chapter six

# Ethics and professional responsibility

The standards by which you conduct yourself and your enterprise are the key pillar in your engineering leadership career. Some of this will be accumulated during your education; however, your ethics on the job will bloom into maturity after you have joined the working economy. It is actual engineering practice that you will have the opportunity of trial-and-error learning as a and experiences that allow you to fine tune your professional acumen and leadership skills to effectively manage a team. These skills should also grow as you advance in your career such that you are prepared to lead larger groups and even an entire organizations. Each company will have its unique corporate culture, which will include company-wide ethics that you will have the opportunity to align yourself with throughout your career. It is paramount to have a personal foundation based on unquestionable integrity as you navigate throughout your career, particularly given the potential impact of the innovations involved in a career as an engineer, manager, or technical leader.

## Ethics in global perspectives

The global economy is now more interconnected than at any time in history. Daily business is conducted with a firm anywhere on the world as effortlessly as if it is across the street. As a result, we are bound by fewer logistical obstacles than ever before. Concomitant with today's new business environment is the challenge to manage ethics across a wide array of different cultures and their respective customs.

Each new local economy joining the flow of global commerce will bring its unique nuance and ethics to the trading community. The cycle by which each region, nation, or community meets and interacts should be grounded in and guided by the common set of concepts and practices determining fairness, right and wrong, and equal justice, or ethics. As the global economy continues to expand, the importance of personal and professional ethics as well as the resulting impact is amplified. The growing, successful company will have leadership that not only will grasp but also will act on the importance of composing the company's foundation every bit as much on sound ethics as it will on the excellence of its product or service.

A sound, substantive viewpoint on the application of your company's ethical framework will include an understanding of both the technological and the cultural impacts of the issues your firm faces. It is a key fact that the far-reaching opportunities resulting from the technology revolution of the past few decades also bring with them great responsibilities for the engineering leaders and their teams. It remains as true today as it was in 1947 at Bell Labs upon the invention of the first transistor; the endeavor to conceive a new breakthrough should be considered not only as "Can we do it?" but also as "Should we?" One effective way of maintaining and building the ethical framework against which these questions can be considered is to measure the present-day ethics against the conventional ethical wisdom of earlier eras. Yesterday's principles and values can be illuminating when compared to and contrasted against other eras, especially today's. The core principles upon which an ethical framework is built are both timeless and universal.

## *Contemporary ethics*

Present-day business ethics are susceptible to the ever-increasing rate of change as are the pace of development of new products and the evolution of new policies addressing the countless social issues that are regularly the subject of Monday-morning water cooler chats. Your individual ethics as an engineering leader are as likely to be influenced, if indirectly, by issues presently affecting populations from the other side of the world as they are by your own family and community beliefs with which you grew up. However, that does not mean that today's ethics are a significant departure from the conduct of business over recent generations. As we redefine ethics for in the present technical environments, the core and guiding tenets of our ethical foundation will remain a combination of truth, integrity, concern for society and equal justice aligned with effective problem-solving business principles that are the foundation of the engineering profession.

"Hold paramount the safety, health, and welfare of the public"—with this statement of ethical foundation began the first "Code of Ethics for Engineers" of the National Society of Professional Engineers (NSPE), which was originally composed in 1946. The code provides NSPE member engineers a framework for ethical judgment. It also outlines the rights, duties, and obligations of practicing members. The NSPE itself was founded in 1934 in New York City as an organization dedicated to any and all nontechnical interests and issues of all licensed engineers. Over the decades, the NSPE has evolved to encompass the full spectrum of topics and needs of today's engineers. The "Code of Ethics for Engineers" has been updated approximately once each decade, as the inexorable forward march of progress has presented new needs and opportunities for each new generation of engineers and engineering leaders. The Code's

Preamble sets out the vision and mission of the NPSE from the first days through today:

> Engineering is an important and learned profession. As members of this profession, engineers are expected to exhibit the highest standards of honesty and integrity. Engineering has a direct and vital impact on the quality of life for all people. Accordingly, the services provided by engineers require honesty, impartiality, fairness, and equity, and must be dedicated to the protection of the public health, safety, and welfare. Engineers must perform under a standard of professional behavior that requires adherence to the highest principles of ethical conduct. [1]

## *Ethics and professional responsibility*

As a technical professional, ethics infiltrates every aspect of our daily activities. The ethics by which we govern ourselves will be revealed in our decision-making. When considering the perspectives on ethical decision-making, five general approaches can be considered. They are ethical relativism, utilitarianism, universalism, human rights, and justice (see Table 6.1).

***Table 6.1*** Summary of five ethical decision-making principles

| Belief systems | Source of moral authority |
|---|---|
| Ethical Relativism (self-interest) | Moral authority is determined by individual or cultural self-interests, customs, and religious principles. An act is morally right if it serves one's self-interests and needs. |
| Utilitarianism (calculation of cost/benefit) | Moral authority is determined by the consequences of an act: An act is morally right if the net benefits over costs (greatest good) are greatest for the majority (greatest number). |
| Universalism (duty) | Moral authority is determined by the extent that the intention of an act treats all people with respect. It includes the requirement that everyone would (should) act this way in the same circumstances. |
| Rights (individual entitlement) | Moral authority is determined by individual rights guaranteed to all in their pursuit of freedom of speech, choice, happiness, and self-respect. |
| Justice (fairness and equality) | Moral authority is determined by the extent that opportunities, wealth, and burdens are fairly distributed among all. |

*Source:* Gary H. Jones, (n.d.) *Five Ethical Decision-Making Principles (Perspectives)*, available at: http://paws.wcu.edu/gjones/Five_Ethical_Perspectives.html#Summary_of_Five_Ethical_Decision-Making

### *Ethical relativism*

As previously expressed, no universal standards or rules can be used to guide the morality of an act. The logic of ethical relativism extends to cultures: cultural relativism. As the saying goes, "When in Rome, do as the Romans do," what is morally right for one society or culture may not be perceived as right by another.

- **Advantage:** Flexibility. Social norms and values are seen in a cultural context.
- **Business implications:** People doing business in a foreign country are obliged to follow that country's social values, norms, and customs (and laws, of course).
- **Engineering implications:** The design of systems, processes, and technologies must consider the population and the associated societal expectations, practices, and customs.

### *Utilitarianism*

An action is judged right or good by its consequences. The ends of an action justify the means used to reach those ends: "The greatest good for the greatest number."

- **Advantage:** Practical, practicable, and especially useful when resources are fixed or scarce.
- **Business implications:** Useful in business (and government) because resources are usually fixed and the "greatest good" is sometimes objective and quantifiable (able to be calculated numerically). This can facilitate (simplify) decision-making.
- **Engineering implications:** This generally simplifies engineering applications as the calculation or estimation of the impact of an engineering product, practice, or service can usually be performed to assess the potential benefit to individuals, organizations, or communities.

### *Universalism*

A person should choose to act if and only if he or she would be willing to have every person on earth, in that same situation, act exactly that same way. There are no exceptions or qualifications. Also, the action must respect all others, and treat people as ends, not means to an end.

- **Advantage:** The interests of people (as ends) are put first. There are no exceptions, special situations, or shades of meaning (but see "criticisms" below).

- **Business implications:** One only makes decisions as one would like to see all other businesses and cultures make that same decision—no exceptions.
- **Engineering implications:** Similar to business implications as one would only design products, services, or methodologies that are beneficial to all individuals who would be potential users.

### Human rights

Individual rights mean entitlements at birth. These entitlements usually include the right to life, liberty, health, dignity, and choice. These rights are often, although not always, seen as being granted to individuals by God. Rights can override utilitarian principles.

- **Advantage:** Human dignity and individual worth are always protected because they are seen as the greatest good.
- **Business implications:** Businesses tend to operate from a cost/benefit (utilitarian perspective). However, business executives should be aware that in many cases, and in many cultures, individual rights must be taken into consideration.
- **Engineering implications:** In considering this perspective, human rights are often used synonymously with "human access." In most cases, engineers and product designers do not influence cultural issues or human rights in an environment. However these technical professionals do have the responsibility to create products, services, and technologies that would promote universal accessibility, minimize bias in product use and are free from features that may be culturally insensitive.

### Justice

The principle of justice deals with fairness and equality.

- **Advantage:** Benefits and opportunities—as well as burdens—have to be shared equally.
- **Business implications:** More easily codified into regulations.
- **Engineering implications:** Easily applied given the legal, regulatory, and regional statutes that govern the development of technical products, services, and innovations.

## Ethical problem approach

These five approaches propose that as soon as we have determined the facts, we should inquire ourselves five questions while attempting to resolve an ethical issue [2]:

- What benefits and what harms will each course of action produce, and which alternative will lead to the best overall consequences?

- What moral rights do the affected parties have, and which course of action best respects those rights?
- Which course of action treats everyone the same, except where there is a morally justifiable reason not to, and does not, show favoritism or discrimination?
- Which course of action advances the common good?
- Which course of action develops moral virtues?

This technique, of course, does not make available an automatic way out to ethical problems. That is not the objective. This technique is simply meant to provide a useful process to guide you in understanding and recognizing the essential ethical considerations in a situation.

## *Applying the frameworks to cases*

While you are utilizing the frameworks to formulate ethical judgments regarding precise cases, it will be helpful to follow the procedure below.

1. **Recognizing an Ethical Issue**
   One of the most essential things you should do at the commencement of ethical consideration is to establish, to the degree possible, the explicitly ethical portion of the issue at hand. Every now and then what seems to be an ethical argument is indeed a dispute about facts or concepts. For example, some utilitarians might dispute that the death punishment is ethical because it discourages crime and consequently generates the utmost amount of good with the slightest damage. Other utilitarians, nonetheless, may argue that the death punishment does not discourage crime and, as a result, produces more harm than good. The disagreement here is about which facts bicker for the ethics of a specific act, not just over the ethics of specific philosophy. Every utilitarian would put up with by the philosophy of turning out the best with the slightest damage.

2. **Consider the Parties Involved**
   Other significant parts you need to reflect upon are the different individuals and groups who may possibly be influenced by your decision. Mull over who may perhaps be harmed or who may possibly benefit.

3. **Gather All the Relevant Information**
   Before you take any action, it is a good initiative to ensure that you have assembled all the relevant facts; furthermore, ensure that all possible sources of information have been consulted.

4. **Formulate Actions and Consider Alternatives**
   Assess your decision-making choices by putting up the ensuing questions [2]:
   a. Which action will produce the most good and do the least harm? (The Utilitarian Approach)
   b. Which action respects the rights of all who have a stake in the decision? (The Rights Approach)
   c. Which action treats people equally or proportionately? (The Justice Approach)
   d. Which action serves the community as whole, not just some members? (The Common Good Approach)
   e. Which action leads me to act as the sort of person I should be? (The Virtue Approach)

5. **Make a Decision and Consider It**
   After investigating all the probable actions, which action you find excellently tackles the circumstances? How do you feel about your choice?

6. **Act**
   Various ethical circumstances are rough for the reason that you can never possess all of the facts. Nonetheless so, you have got to regularly take action.

7. **Reflect on the Outcome**
   What were the outcomes of your decision? What were the anticipated and unanticipated consequences? Would you alter anything now that you have seen the consequences?

Ethical decision-making requires understanding of the ethical repercussion of problems as well as circumstances. In addition, it calls for putting into practice. Having a framework for ethical decision-making is indispensable and useful for enhancing your own understanding in choices making.

## *How to make ethical decisions*

Behaving in an ethical way is always the right thing to do; however, it's not easy all the time. Regularly meeting the requirements of a lofty standard of conduct is not about straightforward moral and immoral decisions, but choosing the "smaller of two evils." A number of decisions necessitate that you prioritize and pick among competing ethical values and principles.

Making ethical decision is based on key personality principles such as honesty, reverence, accountability, justice, considerate, and good citizenship.

Ethical decisions produce ethical behaviors and present a basis for good business practices.

## *Seven-step guide to ethical decision-making*

1. **State Problem**
   a. For example, "there's something about this decision that makes me uncomfortable" or "do I have a conflict of interest?"

   **Case Study 1: Air Bags**

   SafeComp is a company that, among other things, designs and makes sensing devices for automobile air bags. Bob Baines was hired to work in the quality control department. About six weeks after starting work, he was asked to sign off on a design that he felt very uncertain about. He checked with people involved in the design and found the situation, at best, ambiguous. Bob told his manager that he would not feel right about signing off, and, since he was relatively inexperienced with SafeComp's procedures, asked that he not be required to do this. His manager kept applying pressure. Eventually, Bob decided that he wished neither to violate his principles by doing something that he thought was wrong nor to become involved in a battle in which his career would certainly be a major casualty. He quietly resigned [3].

2. **Check Facts**
   a. Many problems disappear upon closer examination of situation, while others change radically.

3. **Identify Relevant Factors**
   a. For example, persons involved, laws, professional code, other practical constraints (e.g., under $200).

   **Case Study 2: Using Other People's Software**

   Jim Warren was a senior software systems expert, hired by NewSoft, a start-up company, to help in the development of a new product. He soon learned that the product was based on proprietary software for which NewSoft did not have a license. Jim assumed that this was some sort of mistake and spoke to the company president about the matter. He was assured that the situation would be rectified. But time passed and nothing happened except that Jim found other instances of the same practice. Repeated efforts to get NewSoft to legalize its operations failed, and Jim, after threatening to notify the victimized companies, was discharged.

   Law enforcement officials were brought into the picture and lawyers on all sides began negotiating. At this date, it is not clear whether criminal charges will be filed. There appears to be a strong possibility of some sort of out-of-court settlement among the companies involved. We don't know how this will ultimately affect Jim Warren [3].

4. **Develop List of Options**
   a. Be imaginative, try to avoid "dilemma"; not "yes" or "no" but whom to go to, what to say.

**Case Study 3: Not Lighting Up**

Will Morgan, a licensed electrical engineer, worked for a state university on construction and renovation projects. His immediate manager was an architect, and next in the chain of command was an administrator, John Tight, a man with no technical background. Tight, without talking to the engineers, often produced estimates on project costs that he passed on to higher university officials. In those cases, not infrequent, where it became evident that actual costs were going to exceed his estimates, he would pressure the engineers to cut corners.

One such occasion involved the renovation of a warehouse to convert some storage space into office space. Among the specifications detailed by Morgan was the installation of emergency exit lights and a fire detection system. These were mandated by the building code. As part of his effort to bring the actual costs closer to his unrealistic estimate, Tight insisted that the specifications for these safety features be deleted.

Will strongly objected on obvious grounds. When he refused to yield, Tight brought charges against him, claiming that he was a disruptive influence. Although his immediate superior, the architect, did not support these charges, he did not fight for Morgan, who was ultimately dismissed by the university. Morgan is now suing for wrongful discharge.

A related issue in this case is that Tight was designating unlicensed people to modify electrical designs submitted by Morgan. This constitutes another improper and indeed illegal act [3].

5. **Test Options**
   Use such tests as the following:
   a. *Harm test*: Does this option do less harm than alternatives?
   b. *Publicity test*: Would I want my choice of this option published in the newspaper?
   c. *Defensibility test*: Could I defend choice of option before congressional committee or committee of peers?
   d. *Reversibility test*: Would I still think choice of this option good if I were adversely affected by it?

e. *Colleague test*: What do my colleagues say when I describe my problem and suggest this option is my solution?
f. *Professional test*: What might my profession's governing body for ethics committee say about this option?
g. *Organization test*: What does the company's ethics officer or legal counsel say about this?

**Case Study 4: Intensive Care**

George Ames, a young software engineer worked for a hospital computer department. He was assigned to work with the people in the intensive care unit (ICU). The computer group was working on the interface between a piece of commercial data processing software and various units in the ICU, including real-time patient monitoring devices.

From the manager down, the computer group was not technically up to the mark in experience or in education. They were falling significantly behind schedule. George learned that they were seriously considering cutting back on testing in order to close the schedule gap. Appalled at this idea, George argued strongly against it. In this case, his arguments had some effect, but he was nevertheless given the clear impression that his prospects with this organization were now significantly impaired. Apparently, part of the problem had to do with reluctance on the part of higher management to clash with the physician who headed the computer group. George felt that the basic problem was incompetence, and he did not see how he could be effective on his own in combating it. About six months later, he resigned [3].

6. **Make a Choice Based on Steps 1–5**

7. **Review Steps 1–6**
   What could you do to make it less likely that you would have to make such a decision again? [4]
   a. Are there any cautions you can take as an individual (and announce your policy on question, job change, etc.)?
   b. Is there any way to have more support next time?
   c. Is there any way to change the organization (e.g. suggest policy change at next departmental meeting)?

## *Application of engineering ethics*

The Code of Ethics is a basic guide for professional conduct and imposes duties on practitioners, with respect to:

- Society
- Employers

- Clients
- Colleagues, including employees and subordinates
- The engineering profession
- Himself/herself

Section 77 of Regulation 941 of the Code of Ethics of Professional Engineers [5] states:

> It is the duty of a practitioner to the public, to the practitioner's employer, to the practitioner's clients, to other licensed engineers of the practitioner's profession, and to the practitioner to act at all times with,
>
> 1. Fairness and loyalty to the practitioner's associates, employers, clients, subordinates and employees;
> 2. Fidelity to public needs;
> 3. Devotion to high ideals of personal honor and professional integrity;
> 4. Knowledge of developments in the area of professional engineering relevant to any services that are undertaken; and
> 5. Competence in the performance of any professional engineering services that is undertaken.

Based on the Code of Ethics as listed above, professional engineers have without doubt an unmistakably distinct obligation to the society, which is to consider the responsibility to public well-being as supreme, exceeding their duties to customers or employers. Your responsibilities to employers' include performing professional responsibilities as loyal representatives or trustees of the organizations. At the same time the utmost concern and care must be exercised regarding confidential customer information. Finally, it is imperative that any potential conflicts of interest be revealed and appropriate measures taken to remove you from these environments where this conflict of interests may exist. Your responsibility to customers means that professional engineers have to divulge without delay any direct or circuitous interest that may prejudice or seems to prejudice your professional decision.

Given the entrepreneurial opportunities available to technical professionals on a part time basis, it is important to manage these situations when also working as a full time employee for another organization. While most companies have formal Conflict of Interest protocols, it is an individual's personal responsibility to ensure that this is addressed. Thus, if you are employed as a member of a team in an organization but you also have external engineering relate work, this information should be disclosed to your employer and to those you interact with as an entrepreneur/individual consultant. As partners and administrators, professional engineers are mandated to collaborate on project work and must not evaluate the work of other professional engineers who are working for the same company without the other's knowledge, and must not spitefully damage the repute or business of other practitioners. Professional engineers are duty-bound to confer appropriate credit for engineering work, sustain the code of sufficient reward for engineering work, and broaden the value of the profession in the course of the exchange of engineering information and knowledge.

## *IEEE code of ethics*

The following is adapted from the Institute of Electrical and Electronics Engineers (the world's largest professional association for the advancement of technology) Policies, Section 7 - Professional Activities (Part A - IEEE Policies).

### *7.8 IEEE code of ethics*

> We, the members of the IEEE, in recognition of the importance of our technologies in affecting the quality of life throughout the world, and in accepting a personal obligation to our profession, its members and the communities we serve, do hereby commit ourselves to the highest ethical and professional conduct and agree [6]:
>
> 1. To accept responsibility in making decisions consistent with the safety, health, and welfare of the public, and to disclose promptly factors that might endanger the public or the environment;
> 2. To avoid real or perceived conflicts of interest whenever possible, and to disclose them to affected parties when they do exist;
> 3. To be honest and realistic in stating claims or estimates based on available data;
> 4. To reject bribery in all its forms;
> 5. To improve the understanding of technology; it's appropriate application, and potential consequences;
> 6. To maintain and improve our technical competence and to undertake technological tasks for

others only if qualified by training or experience, or after full disclosure of pertinent limitations;

7. To seek, accept, and offer honest criticism of technical work, to acknowledge and correct errors, and to credit properly the contributions of others;
8. To treat fairly all persons and to not engage in acts of discrimination based on race, religion, gender, disability, age, national origin, sexual orientation, gender identity, or gender expression;
9. To avoid injuring others, their property, reputation, or employment by false or malicious action;
10. To assist colleagues and co-workers in their professional development and to support them in following this code of ethics.

Ethical thinking comprises the complicated procedure employed to reflect on the effect of our actions on the persons or organization we serve. Despite the fact that most decisions are routine, we can suddenly face an ethical dilemma when extraordinary circumstances happen unexpectedly for which an instantaneous answer is needed.

The basis of making ethical decision involves choice and balance; it is a guide to dispose of wrong choices in support of right ones. Consequently, in making ethical decisions, one of the foremost questions to ask yourself is, "what would a rational man do in this circumstance?" For tougher decisions, you may find the three rules of management by Hojnacki [7] useful:

1. **The Rule of Private Gain**
   If you are the only one personally gaining from the situation, is it at the expense of another? If so, you may benefit from questioning your ethics before the decision.

2. **If Everyone Does It**
   Who would be hurt? What would the world be like? These questions can help identify unethical behavior.

3. **Benefits versus Burden**
   If benefits do result, do they outweigh the burden?

Once people work personally together on a task, they have a tendency to take on the core values of the group. Individuals in a group frequently concede their own ethics in support of those held by the group. As a result, groups should utilize the three rules of management to evaluate whether their organizational decisions are ethical. Research has shown that group dynamics directly impact organizational success as well as standards of behavior within an organization, particularly as it relates to factors impacting the framework of profit and honesty. Thus it is essential that the group understand and conceptualize the impact of their decisions on others and the organization [8].

To be ethically successful, it is vital that you comprehend and revere how values impact your social environment. Recognizing that the way individuals perceive others around them is a significant factor in group dynamics, organizations often create rules of ethical behavior to influence this perception and create an organizational culture designed to have a positive impact on others. Organizations that analyze power and responsibility, and review their ethical decisions frequently, develop employees who function with honesty and integrity and serve their institution and community [8].

## Engineering ethics in practice

Professional engineers work to improve the well-being, health, and safety of each and every one while paying appropriate regard to the surroundings and the sustainability of resources. They have made individual and professional pledge to improve the welfare of society through the development of knowledge and the administration of innovative teams.

Four fundamental principles should guide an engineer in achieving the high ideals of professional life. These express the beliefs and values of the profession and are amplified below.

### Accuracy and rigor

As a professional engineer, you have the responsibility to guarantee that you obtain and utilize wisely and faithfully the knowledge that is applicable to the engineering skills needed in your work and in the service of others. You should [9]:

- Always act with care and competence
- Perform services only in areas of current competence
- Keep their knowledge and skills up to date and assist in the development of engineering knowledge and skills in others
- Not knowingly mislead or allow others to be misled about engineering matters

- Present and review engineering evidence, theory, and interpretation honestly, accurately, and without bias
- Identify and evaluate and, where possible, quantify risks

---

**Case study: An ethics dilemma**

Acting with care and competence: Professional engineers should "always act with care and competence."

**DILEMMA**

You have reason to believe that a colleague, and friend, is not taking sufficient care in the execution of his role of examining engineer, and that this breach of duty may be impacting the safety of rail bridges. You know that you and he both have a duty to ensure that work is conducted with care and competence, but you must decide firstly what such a duty requires in this case and secondly whether any breach is the fault of your colleague or the system in which you are both working. On top of this, you must also contend with conflicts that might arise between your professional duties and your loyalty to your friend. Should you give your friend the benefit of the doubt, report him to superiors, or raise the more general issue of how the system works with your managers?

*Source:* Adapted from The Royal Academy of Engineering and the Engineering Council, *Engineering Ethics in Practice: A Guide for Engineers*, 2011. Available at: http://www.raeng.org.uk/publications/other/engineering-ethics-in-practice-shorter

---

## *Honesty and integrity*

As a professional engineer, you must assume the maximum principles of professional behavior, frankness, equality, and honesty. You should [9]:

- Be alert to the ways in which their work might affect others and duly respect the rights and reputations of other parties.
- Avoid deceptive acts, take steps to prevent corrupt practices or professional misconduct, and declare conflicts of interest.
- Reject bribery or improper influence.
- Act for each employer or client in a reliable and trustworthy manner.

The following box gives a brief engineering example relating to honesty and integrity.

**Case study: An ethics dilemma**

Faisal is a technician working on the central heating system for a building which is occupied by a large financial services company. One day, while carrying out maintenance work in one of the building's corridors, he overhears two executives talking about a debt crisis at the company, something which has not yet been communicated to the public. Later, Faisal's friend, who owns shares in the company, asks him if he knows anything about the company's financial health. Should Faisal warn his friend about what he has heard?

*Source:* Adapted from The Royal Academy of Engineering and the Engineering Council, *Engineering Ethics in Practice: A Guide for Engineers*, 2011. Available at: http://www.raeng.org.uk/publications/other/engineering-ethics-in-practice-shorter

## *Respect for life, law, and the public good*

Professional engineers ought to give appropriate credence to all applicable law, information and published regulations, and the wider public interest. They should [9]:

- Ensure that all work is lawful and justified
- Minimize and justify any adverse effect on society or on the natural environment for their own and succeeding generations
- Take due account of the limited availability of natural and human resources
- Hold paramount the health and safety of others
- Act honorably, responsibly, and lawfully, and uphold the reputation, standing, and dignity of the profession.

Minimizing and justifying adverse effects: Professional engineers should "minimize and justify any adverse effect on society or on the natural environment for their own and succeeding generations."

**DILEMMA**

You are a self-employed engineering consultant. You have been employed to produce an environmental impact statement for a new road tunnel on behalf of the construction company proposing the tunnel. It has been made clear to you that the expectation of your client is that the statement will not find significant environmental problems with the project.

However, you are concerned that if you produce a report that meets these expectations, it will not fully represent the adverse effects of the project and could lead to the project proceeding even though its benefits do not outweigh the environmental damage it will cause. How should you go about completing the environmental impact statement? Should you aim to meet their expectations, adapting the methodology to get the desired results; warn them that the report may highlight problems; or simply produce the most honest, accurate report that you can?

*Source:* Adapted from The Royal Academy of Engineering and the Engineering Council, *Engineering Ethics in Practice: A Guide for Engineers*, 2011. Available at: http://www.raeng.org.uk/publications/other/engineering-ethics-in-practice-shorter

---

## *Responsible leadership: Listening and informing*

Professional engineers must have craving to high standards of management in the utilization and organization of technology. They hold a privileged and trust position in society; consequently, they are expected to display that they are in quest of serving the broader society and to be susceptible to public concern. They should [9]:

- Be aware of the issues that engineering and technology raise for society, and listen to the aspirations and concerns of others
- Actively promote public awareness and understanding of the impact and benefits of engineering achievements
- Be objective and truthful in any statement made in their professional capacity.

---

**Case study: Truth and objectivity**

Truth and objectivity: "Be objective and truthful in any statement made in their professional capacity."

**DILEMMA**

You work as both an independent consultant and as a sales rep for Spectrup, a radio broadcast equipment manufacturer. Given that you are called upon to recommend appropriate equipment to clients in your role as a consultant, you recognize that you are faced with a conflict of interests between your two jobs. As a consultant, you should be impartial, but as an employee of Spectrup you should promote their products where possible. Is it possible to separate your roles to avoid the conflict? And if it is, how do you convince people that you can maintain this separation?

*Source:* Adapted from The Royal Academy of Engineering and the Engineering Council, *Engineering Ethics in Practice: A Guide for Engineers*, 2011. Available at: http://www.raeng.org.uk/publications/other/engineering-ethics-in-practice-shorter

## *Conclusion*

There are many varieties and complexity of ethical issues faced by engineers. These few case studies were adapted for this book although there are countless other cases that could have been used, from all areas of engineering book. One of the purposes of this book was to express the need for engineers to connect with ethical issues in their work, and to illustrate that by unraveling these issues it is possible to make out obvious paths ahead, and not just a coppice of contradictory opinions.

Most importantly, there is an excellent deal of further information obtainable concerning ethics, as well as more case studies, examination of news events, and other resources connected to ethics in the engineering profession.

Next, the knowledge from the case studies in this book can be applied to ethical issues faced in everyday practice.

Engineering is a wide-ranging discipline and the case studies here do not cover all of the ethical issues that an engineer may face. Nevertheless, ethical dilemmas relatively different from those adapted in this book can profit from being approached in a similar method.

## *Chapter summary*

Within all disciplines of engineering, there are several cases in which it is essential to apply diverse types of ethics. The two are virtually indivisible. Engineering entails making many hard-hitting decisions that if made wrong may possibly lead to key penalty. For that reason, it is imperative to many things when making these decisions.

Apparently, several choices are made because of individual viewpoint and principles. On the other hand, within engineering, there is a superior code of ethics that must be put first. More distinctively than that, there are codes of ethics for every particular discipline of engineering that must be considered first before making any key decision. After reviewing these codes of ethics carefully and if there is yet an apparent debate, only then it is acceptable to apply individual viewpoint to assist in making the choice.

The importance of these ethics is not merely to attempt to steer clear of corruption within fields of engineering. It is rather to keep a standard among all engineers so that the reputation of the profession may be kept preserved [10]. Engineers have essential jobs that in a number of cases can hold the lives of countless people in their hands. They need to be people that can be depended thoroughly on a task that can have an effect on the well-being of others without any doubt of corruption [10]. Consequently, engineers must follow their codes of ethics to guarantee that any decision they make is for the right reasons, in addition to the fact that it will shield the honor of the profession.

It is most important to recognize that individual values and judgment are received in the ethics of engineering. Nonetheless, it is imperative that it does not account excessively for a great deal of the decision-making process. Samuel Florman talked about this in his article stating that engineers should not "allow their prejudices to overwhelm their professional discernment" [11]. This gives you an idea about how vital it is not to let one's own bias get in the way of the right decision. With that being understood, Florman furthermore discussed that ethics in engineering are not set in stone, and that they change with the society and technology around them [11]. These are all significant things to reflect on when dealing with both ethics and personal beliefs.

---

## Case studies: Ethical dilemmas

### Case scenario 1: Professional responsibility scenarios

**DILEMMA**

You work for a surveillance technology company, developing behavior recognition systems to protect people from the threat of terrorism—a project that you believe in. However, you have misgivings about the company's new venture into developing hidden cameras for individual use, although you recognize that your company is legally entitled to develop these monitoring products. What do you do [12]?

**WHAT SHOULD YOU DO?**

1. You could tell yourself that you are working on a worthwhile job, and it is not your responsibility to address concerns about public policy and the misuse of products.
2. You could leave your job, stating that you have concerns about the work the company is doing.
3. You could continue working on the project, but at the same time speak out publicly about your concerns about the lack of regulation.

4. You could try to persuade your company to work with you to consider the ethical issues, to work with people in the community and with ethics committees. You could encourage them to work towards campaigning for better regulations while, at the same time, developing the new technologies.
5. You could work with professional bodies to explore the ethical issues, and to campaign for better regulations, informing politicians and the public of the technology that is already available and/or is likely to become available in the near future [12].

**SUMMARY**

In this case, you have a conflict between believing in the value of the project that you are working on and concerns about other products manufactured by the company. It may not be appropriate to quit your job over the issue in the first instance, as you could work towards the sorts of social changes that would be necessary to protect people from the misuse of surveillance technologies. Where possible, it would be best if you could work with the company and/or professional bodies in trying to achieve these aims [12].

*Source:* Adapted from *Engineering Ethics in Practice: A Guide for Engineers:* The Royal Academy of Engineering and the Engineering Council, 2011. Available at: http://www.raeng.org.uk/publications/other/engineering-ethics-in-practice-shorter.

---

## Case scenario 2: Structural integrity of the church

**DILEMMA**

Imagine you are the team leader from Bradlet Structural. It is your responsibility as a consultant to give advice on whether you think a building project is a threat to the structural integrity of a local church. By ignoring your advice and claiming that the church is under no threat, the company who engaged your services, STZ, has given information to the public that you feel to be false; about a topic that has the potential to cause harm to people and property. Furthermore, you have substantial evidence that this is the case, gathered by your team in a professional capacity. How should your team act [12]?

**WHAT SHOULD YOU DO?**

1. You could decide to say nothing, given that the information in your possession was gathered whilst your company was employed by STZ, and there is an obligation to be loyal to those who pay for your services.
2. You could inform STZ that you do not agree with their public statements on the matter of the subsidence around the church, and that they should reconsider their position in the light of the information that your team gathered whilst in their employment [12].

**SUMMARY**

In this case, advising the Building Regulations department of the Local Authority appears to be the best option. This department has the powers to stop any work that they deem to be dangerous and ask for modifications to ensure public safety. Going through the official channels means that, as Team Leader for Bradlet Structural, you can discharge your duty of care to the public, while staying mindful of your responsibilities to your employers. However, if you take this route you may have to decide whether your responsibility ends there. What if the Local Authority does not act on the information? Should you then go public [12]?

---

## *Chapter problems: Three leadership dilemmas*

1. Consider the broader trends driving businesses and economies today and discuss their origins.
2. Describe how your ethical foundation can develop during your career, and your growth into an engineering leader.
3. In what ways can you expect to be able to influence your company, your colleagues, and business partners and customers around the world?

---

**Case study: Ethics and integrity**

On December 2, 2015, a county employee and his wife opened fire on a meeting of San Bernadino County Department of Public Health officials, killing 14 people and injuring others. The massacre ended in a police shootout, which killed the attackers and left the FBI with questions regarding the motive behind what they believed was a terrorism-inspired attack.

One clue left at the crime scene was the smartphone of one of the shooters, which happened to be an Apple iPhone.

Throughout the investigation (which is ongoing at the time of this book's publishing), the FBI has been demanding that Apple build technology that will allow the federal government to hack into the shooter's iPhone to mine encrypted data stored in the phone. This leaves Apple with an ethical dilemma: do they create a "back door" to the iPhone, violating their privacy agreement with current customers and threatening the security of every iPhone in circulation, or do they stand by their promise to protect users privacy, and uphold their own values against what they see as "an overreach by the U.S. government," according to Apple CEO Tim Cook. In an open statement, Cook wrote [13]:

> Once created, the technique could be used over and over again, on any number of devices. In the physical world, it would be the equivalent of a master key, capable of opening hundreds of millions of locks—from restaurants and banks to stores and homes. No reasonable person would find that acceptable.
>
> The government is asking Apple to hack our own users and undermine decades of security advancements that protect our customers—including tens of millions of American citizens—from sophisticated hackers and cybercriminals. The same engineers who built strong encryption into the iPhone to protect our users would, ironically, be ordered to weaken those protections and make our users less safe.
>
> [...] The government could extend this breach of privacy and demand that Apple build surveillance

> software to intercept your messages, access your health records or financial data, track your location, or even access your phone's microphone or camera without your knowledge.
>
> Opposing this order is not something we take lightly. We feel we must speak up in the face of what we see as an overreach by the U.S. government.

In response to Cook's leadership and willingness to stand up for his company in the face of unprecedented legal concerns, other Silicon Valley tech leaders—even those in direct competition with Apple—have come to Apple's defense. Google CEO, Sundar Pichai, tweeted: "Forcing companies to enable hacking could compromise users' privacy …. We build secure products to keep your information safe and we give law enforcement access to data based on valid legal orders …. But that's wholly different than requiring companies to enable hacking of customer devices & data. Could be a troubling precedent."

Twitter cofounder and CEO Jack Dorsey also tweeted his support, saying, "We stand with @tim_cook and Apple (and thank him for his leadership)!" and Facebook submitted the following statement to VentureBeat:

> We also appreciate the difficult and essential work of law enforcement to keep people safe. When we receive lawful requests from these authorities we comply. However, we will continue to fight aggressively against requirements for companies to weaken the security of their systems. These demands would create a chilling precedent and obstruct companies' efforts to secure their products.

Furthermore, an industry group called Reform Government Surveillance, which includes AOL, Apple, Dropbox, Evernote, Facebook, Google, LinkedIn, Microsoft, Twitter, and Yahoo, published a statement on February 17, 2016, in which they insist that "technology companies should not be required to build in backdoors to the technologies that keep their users' information secure."

Whether you're on the FBI's side or Apple's, one thing is clear: Apple's leadership has been admirable in the face of an unprecedented ethical dilemma the company is now facing. Tim Cook has upheld his company's values despite outside criticism, and continues to adhere to Apple's moral compass and do right by its customers.

**QUESTIONS**

1. What engineering leadership characteristics were demonstrated in this case study?
2. How were these characteristics applied to deliver the intended impact?
3. What was the technical, societal, or environmental impact due to the application of the leadership principle(s) identified above?

---

---

**Real profiles in engineering leadership**

**Name:** Brian W. Betts

**Current position or field of expertise:** As Vice President Operations Planning & Insights for Walt Disney Parks & Resorts I lead a global team of analytics professionals, including over 100 Industrial Engineers. Having received a BSIE degree from NC State University, I utilize my Industrial Engineering background to plan and develop new theme parks, attractions and experiences around the globe and ensure the efficient operation of existing experiences.

**What I like most about my position:** I work at the *happiest place on earth.* Each day I focus on products and services that improve the experience for Guests visiting Disney Parks & Resorts. I receive great fulfillment from knowing the work we do makes people happy and creates memories that last a lifetime.

I am also surrounded by an incredibly talented team. I draw inspiration and energy from hearing the ideas and ambitions of my team and helping them reach their developmental and career goals. We provide our team the opportunity to rotate into roles supporting new clients every two to three years. Doing so provides the chance to learn new sides of our business and grow their personal network. It also means we never get bored and always have fun.

**Who or what has made a difference in my career:** I have had many great leaders and mentors throughout my career. Two of the most significant were Erin Wallace and Jeff Vahle—both of whom were EVPs with engineering backgrounds at Disney. They nurtured and challenged me at the same time. I was provided opportunities to travel the world and learn about new cultures and businesses. I was trusted with high levels of exposure early in my career and was always encouraged to speak up.

**I felt like quitting when:** It is often tempting to consider new career paths when I see friends and coworkers pursuing entrepreneurial opportunities with the potential of great rewards. However, I have learned that the grass is not always greener and many end up returning to the company. I am part of a company that aligns with my personal values and continues to provide me with opportunities to grow and advance. That is why I have been with The Walt Disney Company for almost 20 years now.

**My strategies for success are:** I am fortunate to have many visionary leaders from which to guide and draw inspiration for my success. One philosophy that I keep top of mind are Walt's four Cs.

> "Somehow I can't believe that there are any heights that can't be scaled by a man who knows the secrets of making dreams come true. This special secret, it seems to me, can be summarized in four Cs. They are curiosity, confidence, courage, and constancy, and the greatest of all is confidence. When you believe in a thing, believe in it all the way, implicitly and unquestionable." [14]

As Walt says, the greatest of all is confidence. Confidence enables us to do anything our heads and hearts can dream.

**I am excited to be working on:** We are currently putting the final touches on Shanghai Disneyland Resort. This resort represents our sixth resort around the world and our first in Mainland China. I had the opportunity to work on the project from the earliest stages and can't believe we are about to welcome our first Guest to SHDR. The park is spectacular and we are proud to be introducing an entirely new generation to the Disney experience.

**My life is:** I live a blessed life and do my best not to take it for granted. I try to give back and make myself present and available to my team at all times. I am constantly adjusting to find the right balance between my personal interests and priorities and professional responsibilities. I always try to preserve time to reflect and think. It is also important to me to remain active in order to blow off steam and have fun in life.

---

## *References*

1. Preamble. *Code of Ethics for Engineers.* National Society of Professional Engineers. NSPE Code of Ethics for Engineers. Available at: https://www.nspe.org/resources/ethics/code-ethics, accessed on October 13, 2016.
2. Brown University. *A Framework for Making Ethical Decisions.* Available at: https://www.brown.edu/academics/science-and-technology-studies/framework-making-ethical-decisions
3. Stephen H. U. (2000). Examples of real world engineering ethics problems. *Unger, Science and Engineering Ethics,* 6, 423–430. Available at: http://www1.cs.columbia.edu/~unger/articles/ethicsCases.html
4. Davis, M. *Ethics and the University.* New York: Routledge, 1999, pp. 166–167.
5. Professional Engineers Ontario. *Code of Ethics of Professional Engineers.* Available at: http://peo.on.ca/index.php?ci_id=1815& la_id=1
6. IEEE. *Code of Ethics.* Available at: http://www.ieee.org/about/corporate/governance/p7-8.html
7. Hojnacki, William. (2004). Three Rules of Management. In *Managerial Decision Making, graduate course conducted in the School of Public and Environmental Affairs,* Indiana University South Bend, Josephson Institute of Ethics.
8. Chmielewski, C. (2004). *The Importance of Values and Culture in Ethical Decision Making.* Available at: http://www.nacada.ksu.edu/Resources/Clearinghouse/View-Articles/Values-and-culture-in-ethical-decision-making.aspx
9. IESIS. *Statement of Ethical Principles.* Available at: http://www.iesis.org/ethical-principles.html
10. National Society of Professional Engineers. *Code of Ethics.* 2007. Available at: http://www.nspe.org/resources/ethics/code-ethics
11. Florman, S. *Engineering Ethics: The Conversation Without End.* National Academy of Engineers, 2002, Washington: DC.
12. The Royal Academy of Engineering and the Engineering Council. *Engineering Ethics in Practice: A Guide for Engineers.* Available at: http://www.raeng.org.uk/ethicsinpractice
13. Cook, T. 2016. *A Message to Our Customers.* Available at: http://www.apple.com/customer-letter/, accessed on October 13, 2016.
14. Walt Disney. *The Four C's: Curiosity, Confidence, Courage, and Constancy.* Available at: http://www.amyreesanderson.com/blog/the-four-cs-curiosity-confidence-courage-and-constancy/#.V__10n1FZgA, accessed on October 13, 2016.

# chapter seven

# Integration and execution

## Putting it all together

An engineering leader must be able to harness technical skills and integrate them with management capabilities to execute as a successful engineering manager. While technical competence is required for engineers it is equally important for them to develop the necessary non-technical skills in order to be effective manager and leaders. These skills for execution can be summarized in four categories: company values and vision, customers and clients, most valuable asset, and communication.

### Seven essential traits of effective managers

> Leadership is the art of getting someone to do something you want done because he wants to do it. [1]

Different from several of the profession paths for engineers, management is only predestined for a small minority. You must have heard of "the Peter Principle"—it is a concept in management theory formulated by Laurence J. Peter in which the selection of a candidate for a position is based on his/her performance in the current role, rather than on abilities relevant to the intended role. Thus, employees only stop being promoted once they can no longer perform effectively, and "managers rise to the level of their incompetence" [2]. Given technical professionals, this can be interpreted to mean that person is promoted to a management position because they are very good in engineering however they lack the interests as well as the necessary management and leadership to excel in this new position. This seems to occur all too frequently in the engineering profession. Good technological engineers are promoted into supervising or leadership function just to discover that they have a preference doing the technical work and are not good at overseeing a large group of people or setting the direction and vision of their firm. So how do you determine early on that a management leadership career path might be right for you one day? Let's look at a few common traits of managers/leaders:

- **Communication**
  A manager who possesses strong communication proficiency is capable of educating as well as listening. Managers who are capable of

communicating successfully can process information and transmit it back to their teams clearly. For you to be an efficient manager, you must be capable of understanding, deciphering, and relating the organization's vision back to your employees in order to uphold productivity. On the other hand, managers who are ineffective communicators will usually miss the crucial points of what they are being told, and consequently will not be capable of identifying the impact on their team, or will be unsuccessful in sharing the vital points with their team.

As a leader you must know how to communicate to your team effectively. You should know how to deliver messages in a way that will produce a long-lasting impact on the people who listen to you. Great leaders will also pay attention to the people who follow them. Through communication, leaders are capable of conveying precise goals and directions to their followers. Leaders must be good listeners. Many managers are too active contemplating about what to say next, that they are unsuccessful in grasping the important feedback and ideas that are given to them. Remember that, to be a successful manager, an engineer must learn how to be a good communicator. Insufficient communication often means that decisions are made without the benefit of adequate information and as a result the outcomes of these decisions produce poor results.

- **Be a Visionary**
  Human beings will only tag along a vision that is completely clear in their minds; for that reason, it is exceptionally significant that you yourself as a leader have a crystal unambiguous vision of what you want to achieve and the course that you want to go. If your vision is the slightest bit blurred in your mind, how would you then anticipate leading people if you are not clear about where you are going yourself? You must be capable of seeing the big picture and subsequently enhance the plan on how to get there; in order to guide people, you have to keep things easy enough for them to take action on and you must be able to convey your vision to them. Always bear in mind that hazy goals will produce hazy results. Always share your vision with your team and let them take ownership in your dream by illustrating to them how it can profit their lives. Keep in mind that people don't follow other people—they follow the vision that the chosen leader is devoted to. Examine a few great leaders all through history and study their life, and you will indisputably see a truly great visionary.

- **Leadership**
  Leadership is a fundamental characteristic that several managers lack notwithstanding their job title. It is a regular practice for organizations to promote employees who accomplish the top individual results; however, from time to time, the greatest salesman doesn't

make the best manager. True leaders are able to inspire trust, offer direction, and entrust responsibility among team members. An engineer who wishes to become an efficient manager must first acquire basic leadership skill.

- **Adaptability**
  The ability to adapt furthermore contributes to the success of a manager. When a manager is capable of adjusting promptly to unanticipated situations, he is able to direct his team to adapt as well. Adaptability in addition entails that a manager can reflect ingeniously and discovers innovative answer to old problems.

- **Relationship Building**
  Successful managers go all-out to build personal relationships with their teams. Employees are additionally prone to surpass expectations when they have confidence in their manager. Once managers create a relationship with employees, it creates trust and employees feel appreciated. Appreciated employees are more eager to get the job done right and apply additional effort when required.

- **Developing Others**

> Before you become a leader, success is all about growing you. When you become a leader, success is all about growing others. [3]

Successful managers should discern when their employees need further improvement and ways to make sure that those developmental opportunities are successful. Developing others entails refining each individual's ability and inspiring them to conduit their talents toward efficient output.

- **Developing Yourself**
  Finally, a successful manager is conscious of his/her own individual development. In order to effectively enhance and direct others, managers must seek improvement in themselves. Managers who are prepared to continue to grow and learn as well as use their natural talents to the best of their ability will be able to influence the same actions in employees.

Effective management encompasses a number of key components and is not easily realized. Organizations must identify the qualities associated with successful management and then promote employees based on those qualities. The top achieving employees do not usually make the best managers; however, employees who naturally exude these seven characteristics are sure to be efficient and successful in management and leadership positions.

This list of characteristics required to be a successful manager is not all encompassing and only taps upon a small number of qualities that are common for a strong manager and leader. For all engineers those are interested in pursuing a career in this path, it is suggested to begin small by initially taking on a project as a project manager or managing a part-time employee or intern. As the saying goes, "Rome wasn't built in a day," taking a small amount of management and leadership knowledge gradually and developing upon that over the course of your career is therefore the best course of action and the finest way to gain knowledge of how to be a successful leader.

## *How can you prepare yourself for a management/leadership position?*

According to Rachel Cantor Fogarty, the following steps can be taken by an engineer to prepare himself/herself for a management/leadership position [4]:

1. Take a business class. Many leaders and managers have a strong foundation in business, strategic planning, financials, and people management skills. You may also consider continuing on to get your master of business administration (MBA).
2. Ask a mentor or an established leader to take you under their wing and train and guide you toward this path. Many of the best leaders have learned from on-the-job training and have been groomed over the years on the job to become a successful leader.
3. Observe and take a lot of notes from the leaders whom you admire. As the saying goes, "imitation is the sincerest form of flattery," in this case the imitation of the traits and skills of a well-respected leader can be a valuable training tool for someone wanting to learn about leadership.
4. In some companies, employees are offered a management or leadership training program. Seek out programs like this and volunteer for as many activities, promotions, or projects as you can to help you to gain experience in this area.

The pathway toward living a life of impact stipulates that you become an outstanding leader. Begin working on developing your leadership qualities as I have delineated here and on no account stop developing on those abilities for the rest of your breathing days. The better a leader you develop into, the more people will pay attention to and follow your message and the greater constructive change you can influence on humanity as part of your legacy.

### Conclusion

Life doesn't compensate people for what they know; it compensates people for what they do and leaders are doers. Leaders are inspired toward taking action and they understand that for a team to come together, it requires moving from potential to action. The choice is yours to make. Despite the fact that making a decision may lead to failure, the inability to make a decision will completely guarantee failure. It is essential to keep in mind that leaders hardly ever have 100 percent of the information required to make an informed decision. Those leaders who wait pending when they have all the information before they act generally act too late. True leaders are instinctive and pay attention to their gut instincts for the reason that it more often than not gives them the right answer when they are in a state of uncertainty.

## As an engineering leader: Being mission-oriented

It has long been said that one cannot get where one wants to go if he/she doesn't know the way. And key to discovering the way from Point A to Point B is knowing the purpose of the journey. As an engineering leader, it will remain vital that you both develop and maintain a mission-centered game plan in each project you undertake. Make sure to understand; this does not mean having to compose a second-by-second plan before spending the first moment on the project's work. You aren't launching the D-Day landing, after all. The objective is to provide a roadmap that will enable individuals and the team keeps the purpose of the project in focus, while at the same time being enabling sound reasoning and judgment that will reveal the next steps to achieve the project mission.

The cornerstone to remaining mission-oriented is the clearly defined mission statement. By now, this phrase has been so overused across the general lexicon that it has likely lost most, if not all, of its rhetorical value and pertinence. But there is a silver lining in this because it means that revisiting this simple yet valuable step may likely be done with fresh perspective as a clean slate. As they say about clichés, they are clichés for a reason, and the valuable exercise of determining a mission statement is every bit as useful today as it has ever been no matter how overused it may seem to have been. Then take the resulting conglomeration of elements of anecdotes and refine it down to a central image, and let yourself and your team conduct that honing with ruthless effectiveness—you can't break it, and you can always re-add anything you omit. When the exercise has been completed, you will have a highly efficient tool which can guide you in your engineering project and aid you in maintaining a sound focus on your project's mission.

After clarification and communication of the mission for an organization, the effective engineering leader must demonstrate a commitment to the team members to enhance the likelihood that they too will become "mission oriented." The Center for Creative Leadership (CCL) has identified three factors that are useful to help achieve this goal. These qualities include developmental assignments, mentoring/coaching, and recognition of employees [5].

## *Developmental assignments*

Developmental assignments and job rotations are influential motivators for employees. When leaders in official management positions offer a special assignment or a job rotation opportunity, it furthers the employee's career development. Such opportunities also result in a better understanding of and contribution to the organization's wider mission by the employee. Intentionally linking the employee with a specific opportunity to have an effect on the business and the continued development of an idea or product produces an obligation and commitment to the outcome for employee. This blend of opportunity and commitment can significantly keep high potentials and high performers from realizing that they are trapped with no alternative. This blend is particularly significant when a small number of real promotions are obtainable [5].

## *Mentoring/coaching*

Providing employees with the opportunity to have a mentor or coach can be a great motivation for them. Employees can be linked with mentors or coaches who can help them recognize new opportunities, become skilled at strategies for advancing, and draw the attention of higher-level management. Mentoring and coaching offer great accompaniment to additional learning opportunity by supporting and growing on their benefits over time. Furthermore, there is proof that the mentors themselves gain from engaging in a mentoring association [6] and leaders can inspire even top individuals in an organization by offering opportunities to mentor subordinate employees. Mentors gain by having improved visibility within the organization, opportunities to outline their leadership and management skills, and an enduring specialized set of connections.

## *Recognition of employees*

One of the simplest ways a leader can inspire others is to identify efforts and contributions [7]. In contrast to pay raises and promotions, a spoken "thank-you" or an e-mail to share the mission impact of a current project taps into the inner motivations that are inborn to the affirmative twist of both mission and career motivation. Leaders who acknowledge mission impact and communication by encouraging the profession directions of others promote an optimistic, inspiring atmosphere.

## Creating an approach that works for your style and your organization

It is time to start out writing your standards and procedures. This is the exercise in which you have the opportunity to carve out your approach and game plan. You will want to make the plan replicable and turn-key simple; the old somewhat simplistic adage "KISS: keep it simple stupid" is apt and useful here. Your whole leadership personality now has the chance to "mix into the cake" the manner in which you measure and track your ethical framework in a manner similar to how a Middle Ages ship captain would keep a watchful eye on his compass and maps. Let your ethics both inform and enrich your plan-making up to and including your overall planning for how you will set up your standards and procedures, or in other words, how you plan your planning.

Plan on making time to take your approach for a test flight before its implementation. Address it like a thought exercise with some "what if/then" trial-and-error cycles. This might not only reveal any gaps or missing bits that could have slipped through the cracks during your composition of the early drafts of your standards and procedures, it will also provide the opportunity to both confirm whether all of the pieces fit and to change the fit of a few pieces even if the fit was already okay to start with.

## Gaining support for your approach

One of the best ways to present and instill your enterprise's ethical foundation is to model it before your team. There are few better ways to accomplish this than by not merely seeking their buy-in of your standards and procedures but to fully invite them into and include them in the process of composing the drafts of your standards and procedures, including the framework of your ethical foundation. It will be your staff, after all, who will be operating and living with the approach and style that you will lead and execute in your enterprise. So who better to substantively partner with than your team? They can be a rich, deep vein of highly valuable insights and effective tricks-of-the-trade that they will have derived from the actual day-to-day efforts in accomplishing your company's goals. As an engineering leader, you will stand to benefit yourself, your team, and your enterprise by composing your standards and procedures with, for, and around the interests of your employees.

## What not to do

Mistakes are nature's way of showing you that you are learning. As a manager, you will make mistakes, but you can avoid common managerial errors by knowing where the common pitfalls are. Thomas Edison once said that it takes 10,000 mistakes to find an answer. Here are some traps that new and experienced managers alike can fall victim to.

- **Not Recognizing Employee Achievements**
  In these days of regular change, downscaling, and greater than before worker ambiguity, discovering ways to identify your workers for the good that they do is more imperative than ever. The major fallacy is that managers don't want to distinguish employees. The majority of managers concur that gratifying employees is imperative; they just aren't certain how to do so and don't take the time or exertion to distinguish their employees. The most useful reward—individual and written recognition from one's manager—doesn't cost anything. Don't be so full of activity that you can't take a minute or two to identify your employees' achievements. Your employees' self-esteem, performance, and devotion will certainly develop as a consequence.
- **Not Making Time for Your Team**
  As a manager or leader, it is simple to get so engrossed in your own workload that you don't make yourself obtainable to your team. Yes, you have task that you need to deliver. However, your people must come first—without you being accessible when they need you, your people won't discern what to do, and they won't have the support and direction that they need to meet their objectives.

  Steer clear of this mistake by clearing out time in your to-do list purposely for your people, and by learning how to listen actively to your team. Expand your emotional acumen so that you can be more conscious of your team and their needs, and have a standard time when your door is always open, so that your people know when they can get your assistance. Immediately you find yourself in a leadership or management position, your team must always come first—this is, at heart, what good leadership is all about.
- **Not Delegating**
  A number of managers don't entrust roles, for the reason that they feel that no one apart from themselves can do major jobs right. This can result in huge bottlenecks around them, and they gradually

become strained and burned out. Entrustment does take a lot of attempt up-front, and it can be difficult to trust your team to do the work correctly. However, unless you delegate tasks, you are never going to have time to focus on the "bigger-view" that most leaders and managers are responsible for. What's more, you'll fail to develop your people so that they can take the pressure off you.

- **Not "Walking the Walk"**
If you make personal telephone calls during work time or speak negatively about your chief executive officer (CEO), can you anticipate people in your team not to do so? The answer is almost certainly "no." As a leader, you should be a role model for your team. This entails that if they have to stay late, you should also stay late to assist them. For example, if your organization has an expectation or rule that no one eats at their desk, you have a responsibility to set the example and enthusiastically uphold this expectation with your behavior. The same is true for your approach—if you are pessimistic a number of times, you can't suppose your people not to be pessimistic. Therefore, bear in mind that your team is watching you all the time. If you want to shape their behavior, start with your own. They'll follow suit.

- **Failing to Communicate**
In several organizations, the majority of employees don't have inkling about what's going on. Information is power, and a number of managers make use of information, especially the control of information to make sure that they are the most well-informed and, consequently, the most important persons in an organization. A number of managers withdraw away from social circumstances and unsurprisingly keep away from communicating with their employees—particularly when the communication is depressing in some way. Others merely don't make attempts to communicate information to their employees on a continuing basis, allowing other, more urgent businesses take priority by purposely "disregarding" to tell their employees.

  The well-being of today's organizations, particularly during times of change, depends on the extensive distribution of information across an organization and the communication that makes possible this distribution to happen. Employees must be empowered with knowledge, resources and authority that allow them to make excellent decisions at all levels of the organization.

- **Not Setting Clear Goals and Expectations**
Does the phrase "rudderless ship" denote something to you? It ought to. Successful performance starts with unambiguous objectives. If you don't set goals with your employees, your

organization frequently has no direction and your employees have hardly any challenges. For that reason, your employees have little impetus to accomplish anything other than show up for work and collect their paychecks. Your employees' goals start with a visualization of where they want to be in the future. Get together with your employees to grow reasonable, achievable goals that guide them in their endeavor to realize the organization's vision. Don't leave your employees in the dark. Assist them to assist you and your organization by setting goals as well working with them to accomplish organizational goals.

## What to do

One of the major challenges of innovation and leading change is presenting ideas in a way that would prevail over people's apprehension and opposition to change. Sometimes, a good idea is not enough [8]. If you crave support for a fresh change, innovation, invention, procedure, or policy, you need to recognize the psychology behind this resistance, that is, the uncertainties, fears, and doubts that prevent people from accepting change.

There are many reasons for which individuals (as well as workers, board members, and others) resist your ideas. In the following, the major four are presented:

1. **Fear of Failure**
   Contemplate on what you are requesting from others. You wish for them to devote time, effort, and perhaps funds into a project whose result (no matter how positive success seems to you) is unsure to them. Agreed that the majority people are risk disinclined, the fear of failure is an understandable basis of worry and opposition.

2. **Fear of the New**
   Generally, a lot of people will not confront a new trend. Nevertheless, when change actually begins and regular customs, budgets, and job processes are noticeably different, people get confrontational fast. Understand that despite the fact that people may say that they desire new, groundbreaking ideas, they may well not be as receptive when dealing with the domino effect.

3. **Turf Paranoia**
   The initiating of new ideas can produce a swing in precedence and resources that will make a number of people feel alienated, undervalued, or anxious. Extensive transformation initiatives make it hard for people to guard old turf and old ways of doing things. New ideas may indicate an encroachment on the position and authority

held previously. You need to take these apprehensions seriously and consider ways to safeguard the standing and resources of major stakeholders while putting into practice your new ideas.

4. **Confusion**
   New ideas are frequently complex and not easily comprehended for the reason that they may possibly necessitate too much attentiveness and energy from the very people you are attempting to bring on board.

## *How to win support for your ideas*

To triumph over these impediments and get people thrilled about your ideas, improvement, or change attempts, you will have to alleviate these apprehensions as you make your pitch. Some guidelines are given here:

1. **Put Emphasis on the Payoff**
   When putting forward a new idea, leaders time and again presuppose that the benefits of their idea are patently obvious. Emphasize on the benefits, and don't take for granted that everybody will have an instant and deep approval of what you are trying to accomplish. If possible, tell your audience a story that illustrates each and every benefits of your idea.
2. **Couch it in their reality**
   You constantly have to reflect on your audience and their world. You could do well explaining your ideas by illustrating how your ideas relate to their lives.
3. **Address the Risks**
   Most often than not, leaders don't deliberate much on the intrinsic risks of their ideas when presenting them to the team. Take the time to classify areas of risk and the approach to ease those risks.
4. **Be Concrete**
   A number of ideas can be rebuffed for the reason that they sprawl or are too all-encompassing. Tie together all the range of your ideas, and concentrate on precise outlay and timelines. Don't permit the excitement and influence of your idea to make you pledge more than you can deliver.
5. **Put Emphasis on Prestige**
   Conscious of the fact that people will habitually resist change for the reason that they fear loss of position and territory, put emphasis on the social prestige and acknowledgment that they will possibly obtain by joining your effort.

## Getting buy-in:

The commitment of interested or affected parties to a decision (often called stakeholders) to 'buy into' the decision, that is, to agree to give it support, often by having been involved in its formulation. [9]

*Buy-in occurs* when an individual sees clearly that all of his needs, interests, desires, and fears have been fully taken into account.

Buy-in also occurs when a person feels integral part of the group and of the decision-making process where he is asked to provide support.

*The tangible benefit* of buy-in is the fact that whoever buys in will also feel responsible for whatever is being created or produced. People who have seriously bought in into an idea will naturally brag about it, will share the idea with others, and will try to find ways to make it better. In other words, people who buy-in into something will become natural evangelists for that idea/project.

Furthermore, a group of people who have bought in into a project or idea are generally excited about it and will do whatever needed to make it succeed, including being much more flexible and willing to adapt to the needs and requirements of the project. Practically, these people will find extra time to go through the actual changes needed to achieve the goal they have bought in, even if these require them to learn new stuff, change habits, and approach certain things in novel and unfamiliar ways.

*Buy-in can* happen as long as the people asked to contribute and support can clearly see a tangible benefit in their lives for taking this course of action.

The key reasons why it is generally quite difficult to get others to buy-in into your project include the following:

- *Being excluded*
- *Not being valued* properly
- *Not having* one's own interests taken seriously into account

- *Fear of being* exploited
- *Lack of detailed* information
- *Not enough* transparency
- *Having one's* opinion not seriously considered
- *Inability to* understand fully the plan/strategy
- *Things moving* too quickly
- *Lack of affinity* with other stakeholders
- *Lack of clear* motivations for proposed action
- *Lack of clear* benefits and rewards

(Adopted and adapted from Robin Good Master New Media, *What We Really Need to Learn to Be Successful in Life—Part IV*. 2014, available at: http://www.masternewmedia.org/what-to-learn-to-be-successful-p4/#ixzz43mGR9Ua1)

## How-to:

*Inviting teammates,* stakeholders, readers, or fans to provide feedback, criticize, or contribute to your work can be very beneficial, as these people can see your project from different and complementary viewpoints. This can help you refine and hone your approach while getting them to feel tangible ownership of the project.

To get buy-in from other people, the first and foremost thing to do is to inform them in a clear manner that will allow them to identify, empathize, and fully understand the following factors:

1. The issue at stake
2. The consequences of not addressing it
3. The motives and benefits of working together to solve it

The best course of action for anyone wanting to get more buy-in (active support) from others should include as many of the following action points as possible [10]:

1. *Reach* out to, re-unite, and get together all who are or may be affected.
2. Ask about their feelings and fears; gather opinions and sensations on the issue at hand.

3. *Identify* exactly what problems they have. Assess the current situation.
4. *Involve* them in looking and suggesting possible alternative solutions.
5. *Highlight* the real, tangible, and personal consequences that the issue/problem at hand can have on them if not properly solved.
6. *Acknowledge* and publicly describe all of the risks and issues that the project may run into.
7. *Anticipate* all possible arguments, questions, and skepticism, and prepare high-quality answers and fact-rich replies to address them.
8. *Address* and answer all possible doubts, questions, and prejudices.
9. *Clearly* illustrate and explain why your solution/proposal is good.
10. *Clarify* what's in it for you. Why you do it and what you are going to get out of it.
11. *Compare* and confront your proposal/solution to what others have done before or to other possible approaches.
12. *Make* your proposed solution feel as real as possible. If it doesn't feel real, it will not be easy to get buy-in.
13. *Help* yourself by using analogies or storytelling to emotionally convey and share the story and motives behind your project/idea.
14. *Leverage* the power of social media to reach and find more people interested in supporting and helping your idea/project grow.
15. *Explain* clearly how everyone can get involved and how they will be rewarded.

One of the best ways to persuade others is by listening to them. [11]

It is no longer enough to come up with a great idea for a product or service; you need to influence people and gain their support. The challenge is doing that in the face of forces arrayed against innovation within an established organization, including inertia, resistance to change, fear of failure, and financial disincentives. Then there is the obstacle of overcoming people's inability to envision something different and to accept it. Fortunately, there are some effective ways to gain people's support for a new idea that they either don't fully comprehend or aren't convinced will fly.

Getting people to buy into your ideas is not always easy. However, with a little persistence, hard work, and following the basic guidelines, you could also convince others to take a chance on your idea.

## *Leadership and project management*

Your leadership as an engineer has a purpose. In other words, there is a reason for the work you produce in leading your enterprise. On the surface, of course, is the profitable accomplishing of your goals—solving problems through the development of newly engineered products, or enhancements to already developed products. But it will be along the path toward the accomplishing of your goals that you will be able to extract value and meaning from your efforts, all of which will be tethered to your ethical framework. You will have the responsibility to consider the context of your ethical opportunities and challenges as you apply them and also maintain them. And that ethical framework will be presented, and indeed modeled, before your team each day in how you lead them with your choices and the style by which you guide them through each new project. This leadership style will create an environment that allows you to focus on the bottom line, manage day-to-day issues impacting your organization while also still keeping a visionary mindset toward the future.

The standards of project management developed by the Project Management Institute (PMI) (2004) include the application of process groups and knowledge areas deemed critical to project success. *A Guide to the Project Management Body of Knowledge* (*PMBOK Guide*) is a book that presents a set of standard terminology and guidelines (a body of knowledge) for project management. The Fifth Edition (2013) is the document resulting from the work overseen by the PMI. Earlier versions were recognized as standards by the American National Standards Institute (ANSI) which assigns standards in the United States and the Institute of Electrical and Electronics Engineers (IEEE).

*Process groups* of project management involve the following [12]:

### *Process groups*

The five *process groups* are:

1. **Initiating**
   Those processes executed to delineate a new project or a new phase of an open project by acquiring approval to start the project or phase.

2. **Planning**
   Those processes needed to set up the range of the project, refine the objectives, and identify the course of action necessary to achieve the objectives that the project was embarked on to accomplish.

3. **Executing**
   Those processes carried out to complete the work identified in the project management plan to gratify the project stipulations.

4. **Monitoring and Controlling**
   Those processes needed to track, evaluate, and control the development and performance of the project; discover any areas in which alteration to the plan is necessary; and begin the related changes.

5. **Closing**
   Those processes carried out to conclude all activities across all process groups to officially close up the project or phase.

## *Knowledge areas*

The 10 *knowledge areas* are:

1. **Project Integration Management**
   This includes the procedures and activities required to recognize, delineate, merge, unify, and organize the different processes and project management actions within the project management process groups.

2. **Project Scope Management**
   This consists of ensuring that the necessary procedures and resources are available to ensure complete coverage of the activities required for comprehensive project completion.

3. **Project Time Management**
   This consists of the procedures necessary to manage the appropriate conclusion of the project.

4. **Project Cost Management**
   This comprises the procedures involved in planning, estimating, budgeting, financing, funding, managing, and controlling costs so that the project can be completed within the approved budget.

5. **Project Quality Management**
   This includes the procedures and actions of the performing organization that establish quality policies, objectives, and responsibilities so that the project will fulfill the needs for which it was embarked on.

6. **Project Human Resource Management**
   This consists of the procedures that categorize, direct, and lead the project team.

7. **Project Communications Management**
   This consists of the procedures that are needed to guarantee appropriate and suitable preparation, collection, manufacture, allocation, storage, retrieval, administration, control, observation, and the eventual disposition of project information.

8. **Project Risk Management**
   This consists of the procedures of carrying out risk management preparation, classification, analysis, response planning, and controlling risk in a project.

9. **Project Procurement Management**
   This includes the procedures necessary to purchase or obtain products, services, or results required from outside the project team.

10. **Project Stakeholders Management**
    This consists of the procedures required to recognize all people or organizations impacted by the project, examining all stakeholder expectations and impact on the project, and enhancing suitable management strategies for successfully engaging stakeholders in project decisions and implementation.

All of the 10 knowledge areas consist of the procedures that need to be accomplished within its discipline in order to accomplish a successful project management. Each of these procedures falls into one of the five process groups, producing a template formation such that every procedure can be linked to one knowledge area and one process group.

Project managers are still held answerable for the outcome notwithstanding the lack of authority [13]. The project manager's leadership plans and emotional qualities are indispensable elements for obtaining belief of his team and to guarantee project accomplishment. Leading a project toward accomplishment necessitates the manager to get the work done by the team members competently and efficiently. It necessitates the person to have a clear-cut vision, precision in reason, realism in scheduling, and the capacity to attract a brilliant and proficient team.

The application of leadership and management in the project execution typically relies on the kind of project and the life cycle stage that the project is in [14]. For large-scale, complex projects that are universal in nature, the standards to be accomplished, the objectives, as well as the deliverables are inhibited by the project time frame, budgets, and the market dynamics. These categories of projects entail huge and disseminated project teams, encompassing members from various fields. In addition, the execution will be multi-phased. In such circumstances, the project accomplishment and business sustenance can be accomplished sorely through an efficient and smart leadership.

Furthermore, the leadership method should be flexible, sharing, and innovative so as to bring about the project successful accomplishment. At the same time, the leader should put emphasis on team building and inspiration so that the different members can work jointly as a team [14].

---

**Case study: Integration and execution relationship diagram**

NASA's Academy of Program/Project and Engineering Leadership (APPEL) is a program set up to build qualified and successful engineering leaders and teams. Within their Web site, there is a multitude of case study examples [15]. One specific example is the Lunar Crater Observation and Sensing Satellite. This case study is about the communication between managements of two teams, Lunar Reconnaissance Orbiter and Northrop Grumman (NG). The two teams worked on building the same spacecraft for NASA, which made them have to work together as one team. The case study looks into the communication and cooperation between the two teams and how the project was a success for both teams in the end.

**LCROSS CASE STUDY**

When NASA announced that the Lunar Reconnaissance Orbiter (LRO) would upgrade from a Delta II to a larger Atlas V launch vehicle, a window of opportunity opened for an additional mission to the moon. The Atlas V offered more capacity than LRO needed, creating space for a secondary payload.

Ames Research Center served as the lead center for LCROSS. Dan Andrews, LCROSS project manager, was charged with assembling a team that could develop a satellite on a shoestring while coordinating its efforts closely with LRO. "It could have been a real recipe for disaster," he said. "There were plenty of reasons why this mission should not have succeeded"* [16].

**THE GOOD-ENOUGH SPACECRAFT**

From Andrews's perspective, the LCROSS spacecraft had to be "faster, *good enough*, and cheaper." He made clear to his team from the beginning that LCROSS was not about maximum performance. "It was about cost containment," Andrews said.

* Adapted from Academy of Program/Project & Engineering Leadership, *Lunar CRater Observation and Sensing Satellite (LCROSS)*, available at: http://appel.nasa.gov/wp-content/uploads/sites/2/2013/04/474589main_LCROSS_case_study_09_23_10.pdf

"LCROSS was not about pushing the technical envelope. It was about keeping it simple—keeping it good enough."

The LCROSS team had 29 months and $79 million to build a Class-D mission spacecraft. Class-D missions, which are permitted medium or significant risk of not achieving mission success, must have low to medium national significance, low to medium complexity, low cost, and a mission lifetime of less than two years. The low-cost, high-risk-tolerance nature of the project led to a design based on heritage hardware, parts from LRO, and commercial off-the-shelf components.

LCROSS's status as a Class-D mission did not preclude the team from practicing risk management. "We were risk tolerant, but that doesn't mean we were risk ignorant," said Jay Jenkins, LCROSS program executive at NASA Headquarters.

"With the LCROSS instrument testing, we shook, cooked, and cooled the mostly commercial off-the-shelf parts that could potentially come loose during launch so that we were likely to have a tough little spacecraft, but we didn't test to failure," said Andrews.

## TEAMWORK

Andrews knew he had to establish trust with the other teams involved in the mission. LRO, based at Goddard Space Flight Center, was understandably cautious about LCROSS hitching a ride to the moon with them. Andrews quickly moved to identify an LCROSS engineer who could take up residence with the LRO team to facilitate quick dialogue and build trust between the two missions. With good lines of communication, the two teams started to view each other as resources and "worked together like a team this agency hasn't seen in a long time," said Andrews. "These good relationships really pay off when things get tough."

The crossover and reuse of hardware between the LCROSS and LRO spacecraft allowed the teams to learn from and with one another. Sometimes they worked in tandem; at other times, one team would be ahead of the other. "There were things that we missed that we either caught later, or missed and LRO caught, and vice versa," said LCROSS Project Systems Engineer Bob Barber.

A good partnership with Northrop Grumman (NG), the spacecraft contractor, was also essential. Neither Andrews nor NG Project Manager Steve Carman had ever managed spacecraft development before, though both had run spaceflight-hardware development projects.

During the first six months, as the project underwent some acquisition-related contractual changes, Andrews and Carman began to develop a mutual trust. "Ultimately, communication was the hallmark of the partnership," said Carman. "The partnership was not something where we said, 'Sign here—we are partners.' It grew out of a relationship. … We showed them as we went along that we were indeed capable of doing this faster than anything we had done here."

For Andrews, trust grew out of a shared understanding of the way both organizations traditionally operated. "We talked plainly about budgets. We talked plainly about the NASA construct, and then we talked plainly about how hard it is to move NG's heavy institution," he said. "I was not holding anything back in terms of what I was sharing with them, and I think that set a tone within NG [so] that they behaved similarly."

By the time of the preliminary design review, a cooperative dynamic had been established that went beyond business as usual. "It was an 'open kimono' type relationship where everything was kind of on the table," said Barber. "We wanted a really open and honest relationship with them." NASA team members took part in NG's risk management boards and were invited to staff meetings.

The relationships didn't end when people left the project. Both NASA and NG experienced turnover, which could have hurt the project. In this case, though, several former team members kept in touch with their successors. "That's when you know a team is more than just coming to work and doing stuff," said Barber.

Against long odds, the project met its cost and schedule constraints and passed its final reviews. It was time for launch. The Atlas V launched LCROSS to the moon on Tuesday, June 18, 2009.

## PROFESSIONAL DEVELOPMENT AS AN ENGINEER OR LEADER

The term "professional development" may different interpretations for various organizations, environment and individuals. Often we include "networking" under the professional development umbrella because building a network of professionals that you can call on for expert advice is important to any business. However, we should think about professional development like a Swiss army knife with several tools attached to it, and networking is only one of those tools. Ultimately, these tools will allow you to develop your career path as an engineer and a leader.

It is important to remember that nobody is in charge of your career but you, and there are several tools in this book that will help you seek out opportunities for yourself so that you can proactively work toward your professional goals. You will make mistakes along the way, but taking risks and learning from those mistakes are crucial to your personal growth and development. How you react to situations will not go unnoticed, and it is important to handle both your mistakes and your successes with grace. Be proactive about asking your boss, manager, or mentor for feedback or constructive criticism so that you can improve your professional skill set and further develop the areas that need improvement as you grow into the engineering leader that you are destined to become in your career.

---

**Real profiles in engineering leadership**

**Name:** Dr. Claudia J. Alexander

**Current position or field of expertise:** U.S. Rosetta Project Scientist (Project Scientist representing the NASA contribution to the International Rosetta Mission)

**What I like most about my position:** I represent the United States of America when I am overseas doing my job. What a responsibility! We want to assure that NASA is a respected and sought-after partner for collaboration in space exploration. I take that role very seriously. In addition, we work as a team to understand the mysteries of comets. Solving science problems is like a puzzle. The challenge is very satisfying.

**Who or what has made a difference in my career:** My faculty advisors have been critical to learning the ropes in my field. In addition, I have had one or two senior people step in at critical places in my career to help steer me, and to help hone skills I already had but wasn't using very well. One was a professor in undergraduate school. Another was someone in the workplace who overheard me yelling at someone over the phone one day!

Finally, I should mention that my parents had high expectations and set high bars for their kids. They sacrificed and made sure that we could be positioned to be successful academically. They insisted that I do something technical for a career, and I was not happy about it for a while, but in the end they were right. For this, I'll always be grateful.

**I felt like quitting when:** Early in my career, I was working 90-hour weeks for about 6 weeks trying to make deliveries on both the Galileo project (exploring Jupiter) and our brand-new internet site (in the 1990s, when the internet was completely undeveloped, we had a grant from NASA to develop one of the first science-learning Web sites). I was supposed to be doing 50 percent on each project, which was 45 hours per project and well over full time! It was exhausting. When one delivery was done and I dropped down to 45 hours on the second project, I was reprimanded by the supervisor of that project for charging them too much. It's discouraging not to have one's work appreciated, or worse, to feel like the extra hours are expected and the supervisors are looking the other way while you overextend yourself. This is nothing new in the workplace, however. And the hard work we put into developing the internet has been rewarded ten thousand-fold in the decades since.

**My strategies for success are:**

- Listen
- Never believe you are the smartest person in the room—someone else may have a great deal to contribute to any discussion
- Think outside of the box
- Be essential in any project you take on
- Pay attention to innovations that are going on outside of your home venue.

**I am excited to be working on:** Exploring a small body like a comet with the Rosetta spacecraft. It is a privilege to work with a team on a flagship-style mission (one of the biggest and most ambitious of space exploration missions).

Learning more about the origin of the solar system.

Creating science-learning books for school kids to pass along everything I've learned about the planets in a (hopefully) fun way. I'm trying to create something for kids to read that I would have wanted to read in third to fourth grade. Something with a few fantasy characters, a little magic, but with the fun of exploring other planets and stars—something I've also learned a little about in my career.

**My life is:** Amazing right now! A project scientist is at the height of her profession, and it's a very exciting role. Writing books (and winning awards as an author!) fulfills a deep desire that I've harbored since girlhood. (I have stories beating their way out of my brain.) It's a huge blessing to be able to address everything I've wanted to do as a professional. And I'm grateful to have been able to see the culmination of a lot of hard work.

**Name:** Lisa Macon, Ph.D.

**Current position or field of expertise:** Dean, Division of Engineering, Computer Programming, and Technology at Valencia College. Advanced degrees in computer science and mathematics.

**What I like most about my position:** I am privileged to lead a diverse group of brilliant learning leaders at Valencia College. We are all highly focused on improving the lives of people within our community through education in high-tech fields. Our team specializes in taking the "diamond in the rough" and through hard work, resilience, dedication, and perseverance, helping that individual to have a successful, high-tech, high-wage career. No kid gloves here—just lots of support and believing in our students and all they can be.

**Who or what has made a difference in my career:** In tenth grade, my computer science teacher, Mr. Acevedo, gave me a B as a semester grade even though my course average was an A. My father requested a parent/teacher conference to find out why. Mr. Acevedo told my dad that he would change the grade, but that he did not want to encourage me to be a computer scientist because this was not a good field for women. With my father's encouragement, I decided right then and there to prove Mr. Acevedo wrong and I have spent most of my life doing just that.

**I felt like quitting when:** I was trying to finish my Ph.D. and simultaneously juggle work, marriage, small children, and my dad's heart surgery. But I hung in there and finished. It was absolutely worth it.

**My strategies for success are:** Never use the negative things going on in your life as excuses. Use them as driving forces. If you can succeed in spite of these things, and not fail because of them, you're pretty awesome, aren't you?

**I am excited to be working on:** Developing innovative strategies to encourage young women to consider fields in engineering, math, and technology. These are great careers for women and I'm not sure that it's clear to young girls how much they can contribute to society and the human spirit by choosing careers in these areas. I'm very excited to see the results of upcoming projects in this arena.

**My life is:** Busy, hectic, and loaded with fulfilling moments that take my breath away. There are not enough hours in the day to do all the things I'd like to do, but I try to dedicate myself to making the world a better place for my children and for all future generations.

---

## References


1. Eisenhower, D.D. *Breakthrough Training*. Available at: https://breakthroughtraining.com/success-articles/8-traits-false-leadership/, accessed on October 14, 2016.
2. Peter, L. J., et al. *The Peter Principle: Why Things Always Go Wrong*. New York: William Morrow and Company, 1969, p. 8.
3. Rachel Cantor Fogarty. 2014. *Six Essential Traits of Effective Managers*. Available at: http://blogs.asce.org/six-essential-traits-of-effective-managers/
4. Welch, J. *Pschology Tody*. Available at: https://www.psychologytoday.com/blog/the-mindful-self-express/201204/six-qualities-leaders-need-be-successful, accessed on October 13, 2016.
5. Center for Creative Leadership and Booz Allen Hamilton. 2011. *Motivated by the Organization's Mission or Their Career? Implications for Leaders in Turbulent Times*. Available at: https://www.boozallen.com/content/dam/boozallen/media/file/Motivated_by_Mission_or_Career.pdf, accessed on October 13, 2016.
6. Lentz, E. & Allen, T. D. The role of mentoring others in the career plateauing phenomenon. *Group & Organization Management*, 34, 358–384, 2009.
7. Luthans, K. Recognition: A powerful, but often overlooked, leadership tool to improve employee performance. *Journal of Leadership Studies*, 7, 31–39, 2000.
8. Bacharach, S. *Great Leaders Know That Good Ideas Aren't Enough*. 2004. Available at: http://www.inc.com/sam-bacharach/great-leaders-know-good-ideas-arent-enough.html, accessed on October 13, 2016.
9. Wikipedia. Buying in. Available at: https://en.wikipedia.org/wiki/Buying_in, accessed on October 13, 2016.
10. Robin Good Master New Media. *What We Really Need to Learn to Be Successful in Life—Part IV*. 2014. Available at: http://www.masternewmedia.org/what-to-learn-to-be-successful-p4/#ixzz43mGR9Ua1, accessed on October 13, 2016.
11. IEEE. *IEEE Guide—Adoption of the Project Management Institute (PMI(R)) Standard: A Guide to the Project Management Body of Knowledge (PMBOK(R) Guide)*. 4th ed. 2011, New York, NY.

12. Montague, J. Frequent, face-to-face conversation key to proactive project management. *Control Engineering*, 47(6), 16, 2000.
13. Williams, P. 2002. FaithWords. *The paradox of power: a transforming view of leadership.*
14. Chittoor, R. *Importance of Leadership for Project Success.* 2012. Available at: http://project-management.com/importance-of-leadership-for-project-success/, accessed on October 13, 2016.
15. NASA. *Case Study Examples.* Available at: http://appel.nasa.gov/knowledge-sharing/case-studies/, accessed on October 13, 2016.
16. Academy of Program/Project & Engineering Leadership. *Lunar CRater Observation and Sensing Satellite (LCROSS).* Available at: http://appel.nasa.gov/wp-content/uploads/sites/2/2013/04/474589main_LCROSS_case_study_09_23_10.pdf, accessed on October 13, 2016.

# Index

## D

## F



## G



## H



## I

## M

## Q



## R



## S